# Avoiding Population Collapse
Illustrated Science Exploration by Rolf A. F. Witzsche

© Text Copyright Rolf A. F. Witzsche 2018
all rights reserved

This book contains the transcript with images of the exploration video with the above title:
see: http://www.ice-age-ahead-iaa.ca/

## Lead in:

*We are the children of a long sequence of ice ages that are far from being over.*

In times between the glaciation periods, populations had expanded, and then collapsed again when the climate conditions became harsh during the glaciation stages.

We are near such a collapse situation in our time. Our World population has expanded explosively during the last 200 years of the present interglacial period. It had expanded from 1 billion in the year 1800, to the more than 7 billion that we have today. This wasn't the result of improved breeding habits.

It resulted from 3 factors coming together, which made this miracle possible. One factor was the dawn of science, which had its root in the development of classical culture. The second factor was the dawn of industrialization and technologies that together with science made improved agriculture possible that enabled evermore people to live. The third factor was the radical improvement of the climate that enabled scientific and technological progress to expand the base for human living almost without bounds.

The up-ramping in solar activity from the early 1700s onward gave us almost 300 years of amazing global warming.

The Little Ice Age if the 1600s had been a scene of population decline by starvation, even cannibalism in some cases to ward off starvation. All this changed when the solar up-ramping began, with which the climate recovered and became warm again. Modern agriculture would not have been possible without the massive solar global warming that had occurred. Most people in the world would not be living if the sun-caused global warming hadn't happened. The vast majority of our food comes from agriculture. When agriculture thrives and expands, which is possible in warm climates, humanity thrives and expands with it.

But now that the solar global warming event is over and the global warming is fading, one of the fundamental supports for our living on this planet is fading with it. In fact, the global warming that rescued us from the Little Ice Age isn't just fading. It is collapsing, and agriculture is beginning to collapse with it. While the collapse is still in the early stages, large-scale crop losses have already been experienced.

Are we facing a large-scale population collapse then, as the result of the climate collapse? Are we facing inevitable massive crop failures in the years ahead? I would answer no.

The answer has to be no, because we have the scientific and technological potential developed to lift agriculture out of its dependence on climate conditions. If it wasn't for this potential, we would be seeing an enormous population collapse in the near future, such as we have never imagined. But, as I said, I don't see this happening. This is so, because we have the power toady to avoid the population collapse that follows collapsing agriculture as we know it, by simply relocating our agriculture into the tropics and into indoor facilities there.

This power, that we have on this scene, is not trivial, and it is real. It is amazing and grand, and worth celebrating. When this creative power is applied, the human scene unfolds into joy, and peace erupts along the way.

## Table of Contents

Humanity has lived on planet Earth for roughly 2.5 million years .................................................. 7

We are the pinnacle of it all, with unsurpassed power ................................................................ 8

Will we live up to who we are ? .................................................................................................. 9

The human journey began when the Ice Ages began ................................................................ 10

Our World population expanded explosively during the last 200 years ..................................... 11

It resulted from 3 factors coming together ................................................................................ 12

The third factor was the radical improvement of the climate .................................................... 13

The up-ramping gave us almost 300 years of global warming ................................................... 14

Now that the solar global warming is fading out ....................................................................... 15

While the collapse is still only in the early stages ....................................................................... 16

Are we facing a large-scale population collapse then ? ............................................................. 17

We have the scientific and technological potential developed .................................................. 18

There is no joy in being dead. .................................................................................................... 19

There is no real alternative to being alive with joy .................................................................... 20

Only 1 of the 8 major human species has survived, and that's us ............................................. 21

Population Bottleneck roughly 150,000 years ago .................................................................... 22

In seaside caves at a place called Pinnacle Point ....................................................................... 23

Devastating impact a major climate collapse can have ............................................................. 24

This minuscule 4 million worldwide, was all that we had after 2.5 million years ...................... 25

What had caused this increase from the 1800s onward ............................................................ 26

The Sun, getting stronger in solar activity, had enabled a renaissance in living ........................ 27

The recovery of the climate had opened up the portal for science ........................................... 28

Now that the climate recovery by the Sun is over ..................................................................... 29

Rate of collapse reflected in the collapsing solar-wind pressure ............................................... 30

The spacecraft measured a 30% collapse per decade ..... 31

The solar-activity collapse tells us ..... 32

Climate collapse will affect agriculture in a big way ..... 33

April 2018 winter-type blizzard, named Xanto, ..... 34

The natural growing window is shrinking ..... 35

Climate collapse will shrink the growing season ..... 36

Nor will the climate collapse affect only Canada ..... 37

Europe is similarly affected ..... 38

The collapse in solar activity is continuing ..... 39

Increasing volumes of solar-cosmic ray flux cause increased cloudiness ..... 40

Increased cloud nucleation also causes faster rainout ..... 41

Even the greenhouse effect is diminishing ..... 42

Climate is now collapsing back to the Little Ice Age ..... 43

We cannot escape or halt the now ongoing climate collapse ..... 44

When we drop below the Little Ice Age level, into glaciation ..... 45

We also have to make do with 80% less rain ..... 46

To build us a new world synonymous of escaping a death sentence ..... 47

We didn't have this power developed in the 1600s ..... 48

Win, and come out richer with a brand new world ..... 49

We see the writing on the wall, and its promise ..... 50

There is joy in building, and no joy in being dead ..... 51

More Illustrated Science Books by Rolf A. F. Witzsche ..... 52

# Humanity has lived on planet Earth for roughly 2.5 million years

Humanity has lived on planet Earth for roughly 2.5 million years. Its path was one of obstacles, trials, and achievements that no lesser species in the history of life has been able to achieve.

# We are the pinnacle of it all, with unsurpassed power

We are the pinnacle of it all, with unsurpassed power to protect our life, and all life. But will we use the power that we have the potential to wield?

Will we live up to who we are ?

Will we live up to who we are, or will we allow us to simply fade into oblivion as if we have never existed? The latter is the current course.

# The human journey began when the Ice Ages began

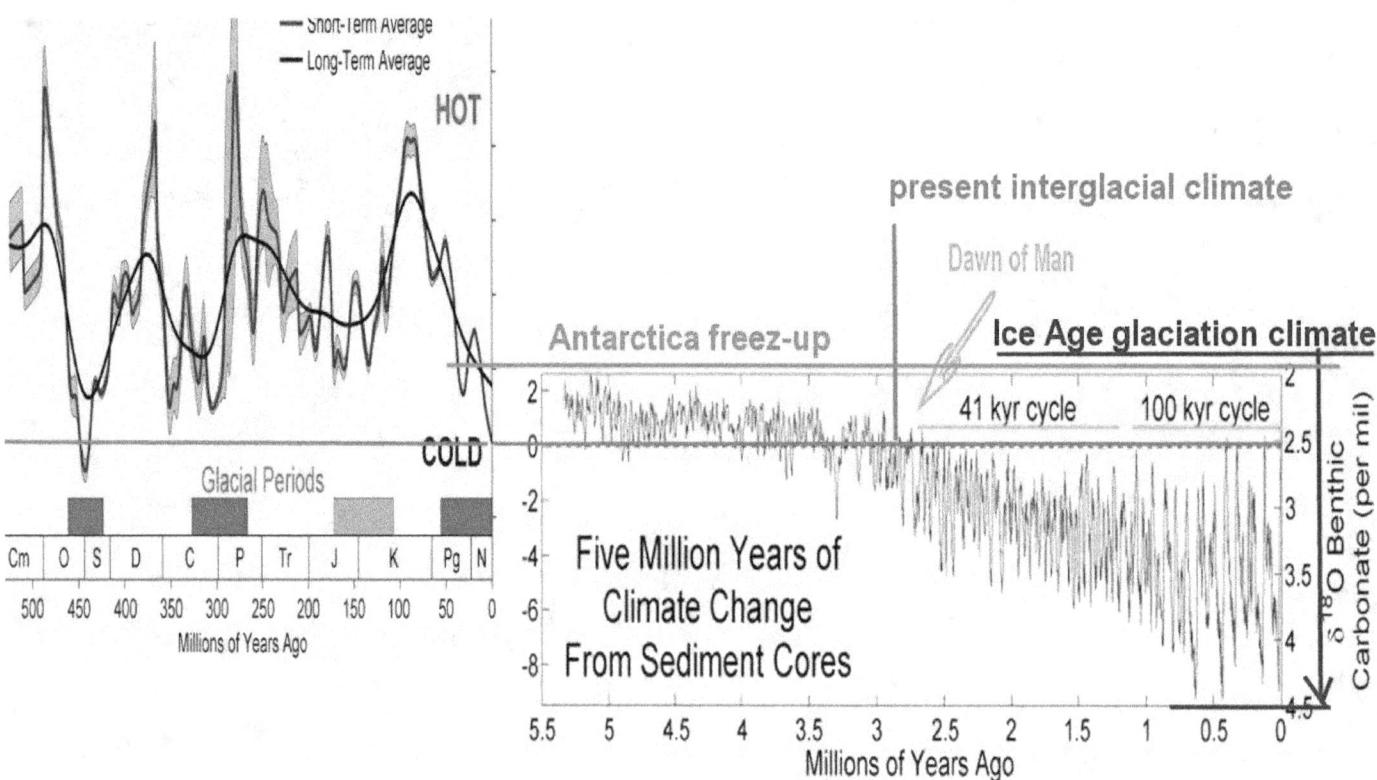

The human journey began when the Ice Ages began. We are the children of a long sequence of ice ages that is far from being over. In times between the glaciation periods, the populations had expanded, and then had collapsed again when the climate conditions became harsh.

# Our World population expanded explosively during the last 200 years

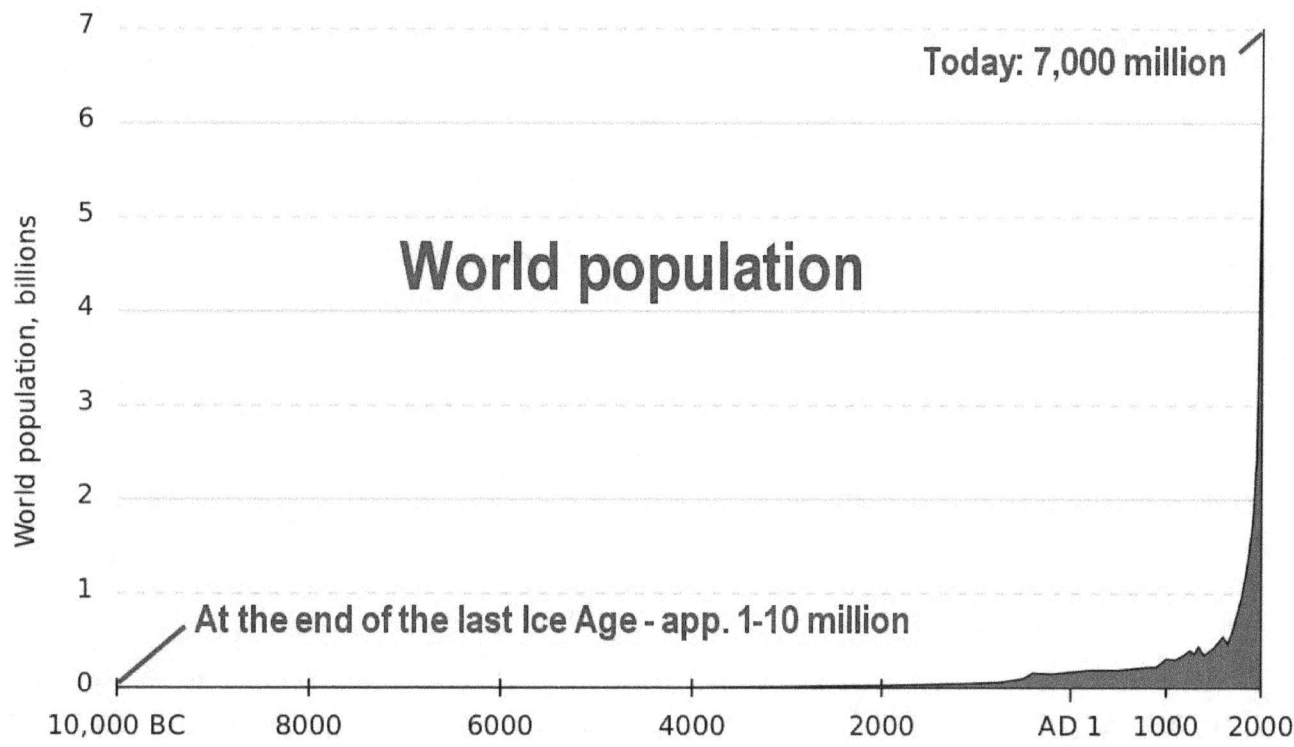

We are near such a situation in our time. Our World population expanded explosively during the last 200 years of the present interglacial period. It had expanded from 1 billion in the year 1800, to the more than 7 billion that we have today. This wasn't the result of improved breeding habits.

# It resulted from 3 factors coming together

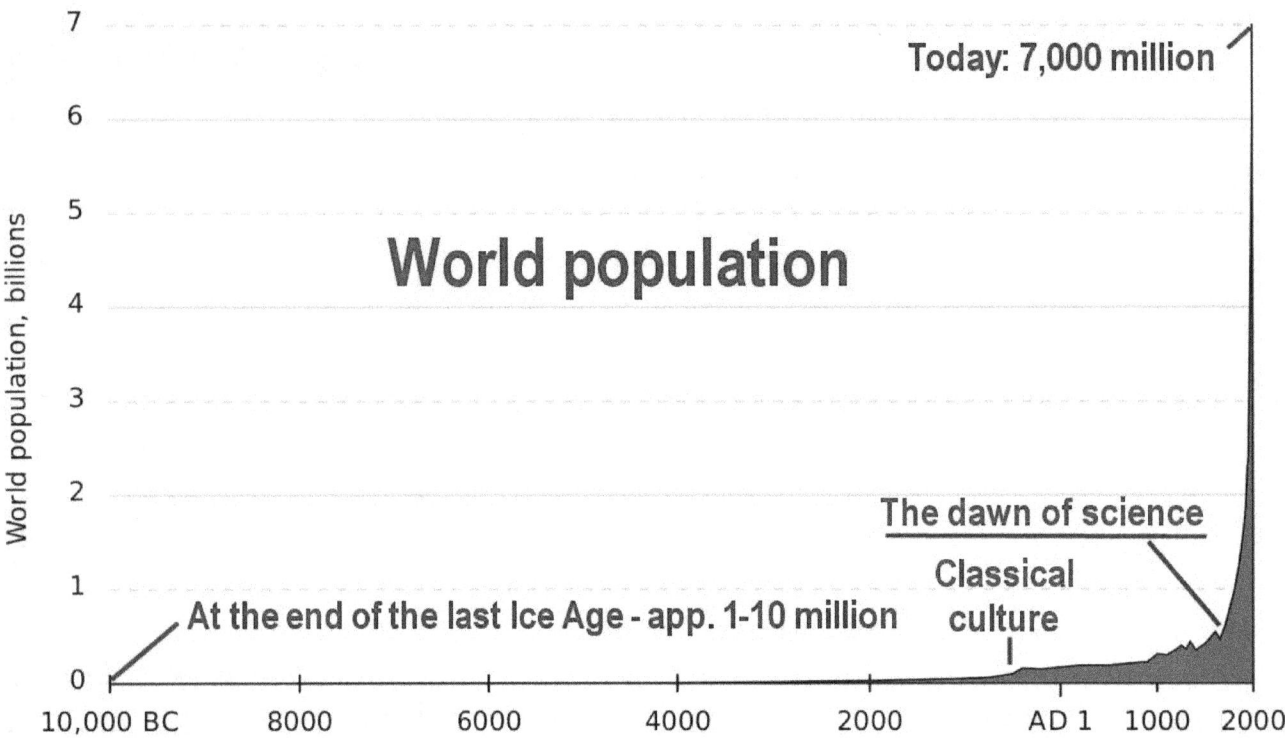

It resulted from 3 factors coming together, which made this miracle possible. One factor was the dawn of science, which had its root in the development of classical culture. The second factor was the dawn of industrialization and technologies that together with science made improved agriculture possible that enabled more people to live.

# The third factor was the radical improvement of the climate

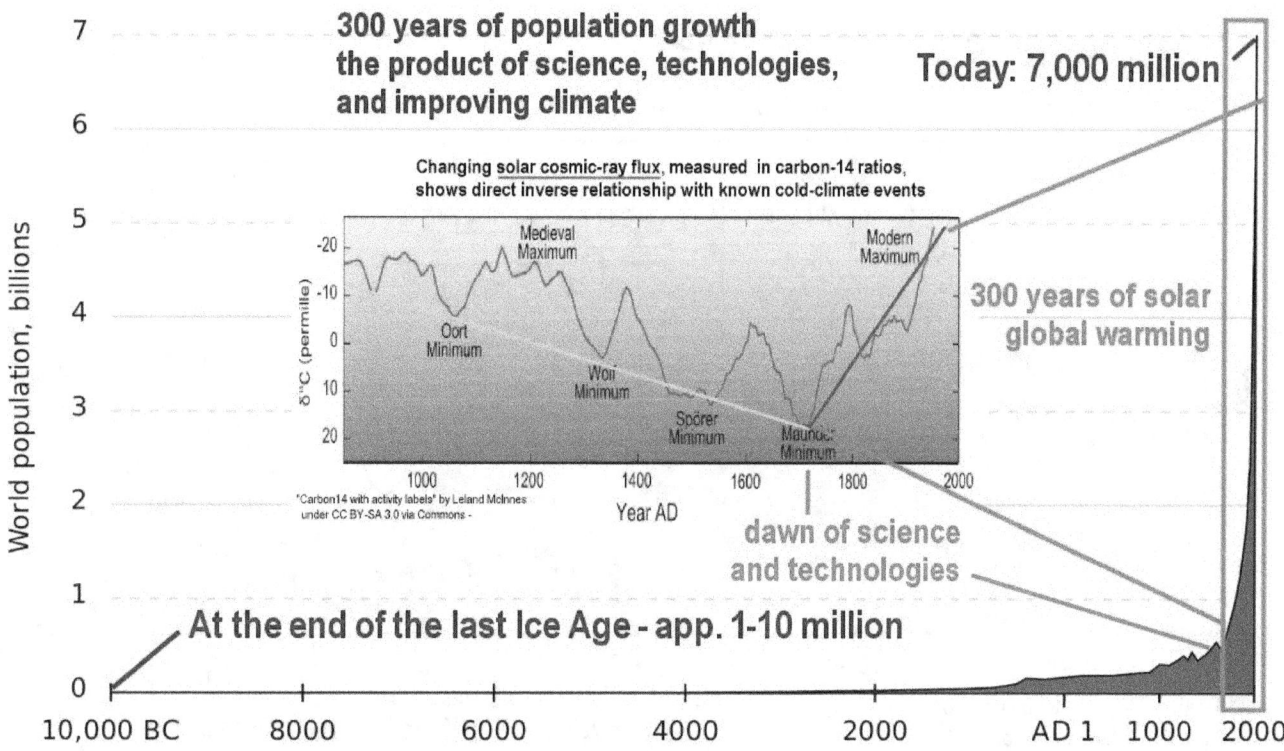

The third factor was the radical improvement of the climate that enabled scientific and technological progress to expand the base for human living almost without bounds.

# The up-ramping gave us almost 300 years of global warming

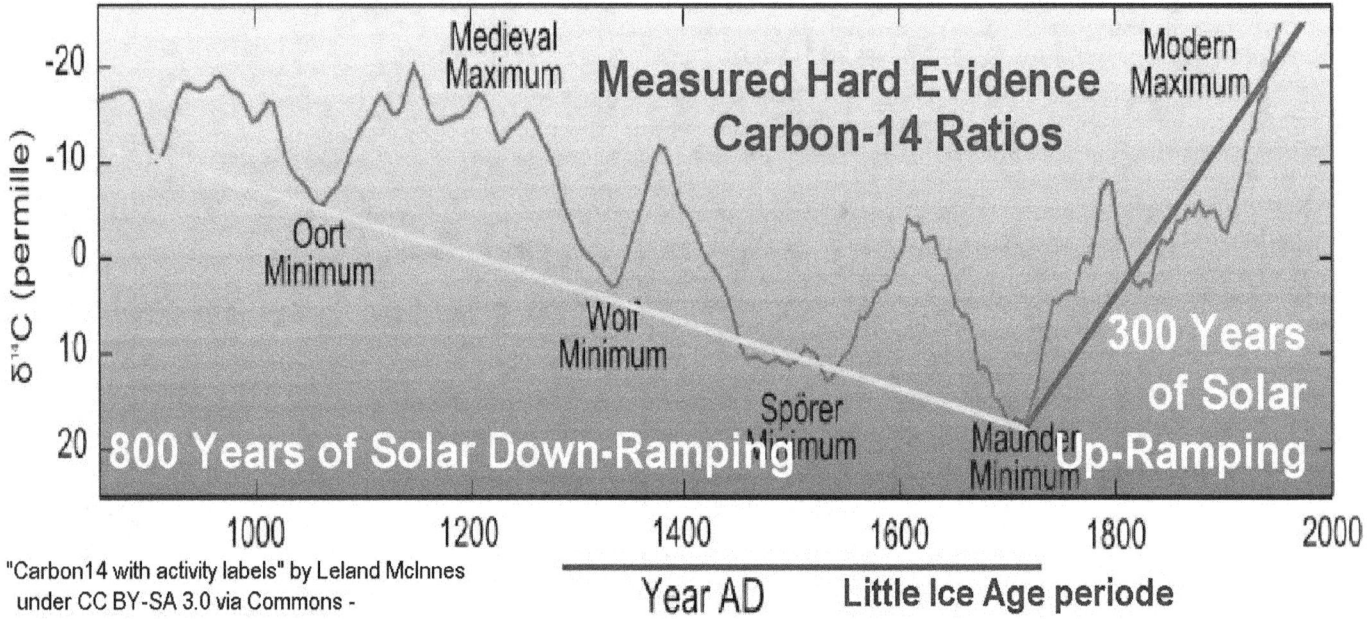

The up-ramping in solar activity from the early 1700s onward gave us almost 300 years of amazing global warming. Modern agriculture would not not have been possible without this amazing solar global warming. Most people in the world would not be living, if that global warming hadn't happened. The vast majority of our food comes from agriculture. When agriculture thrives and expands, which is possible in warm climate, humanity thrives and expands with it.

# Now that the solar global warming is fading out

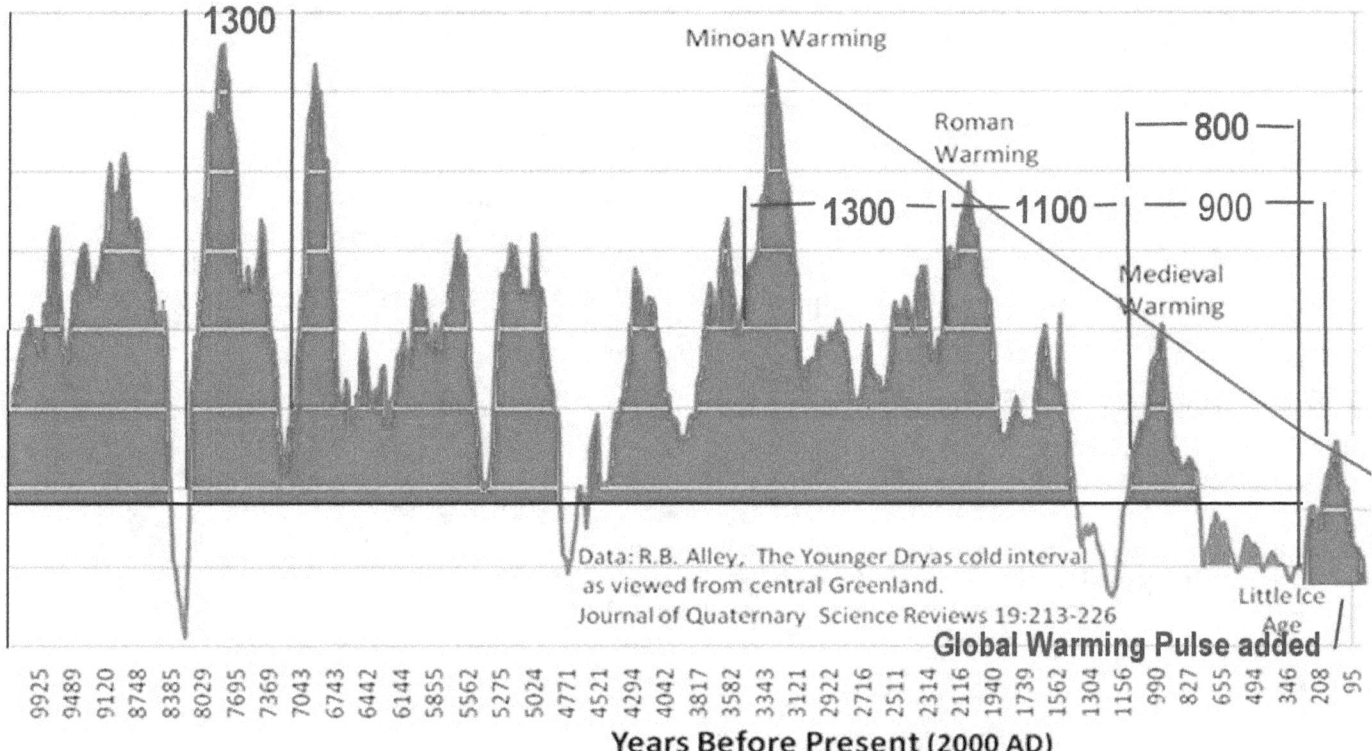

But now that the solar global warming event, the last of the 4 solar warming pulses is fading out, one of the fundamental foundations for our living on this planet is fading out with it. The global warming pulse isn't just fading. It is fast collapsing, and agriculture is beginning to collapse with it.

# While the collapse is still only in the early stages

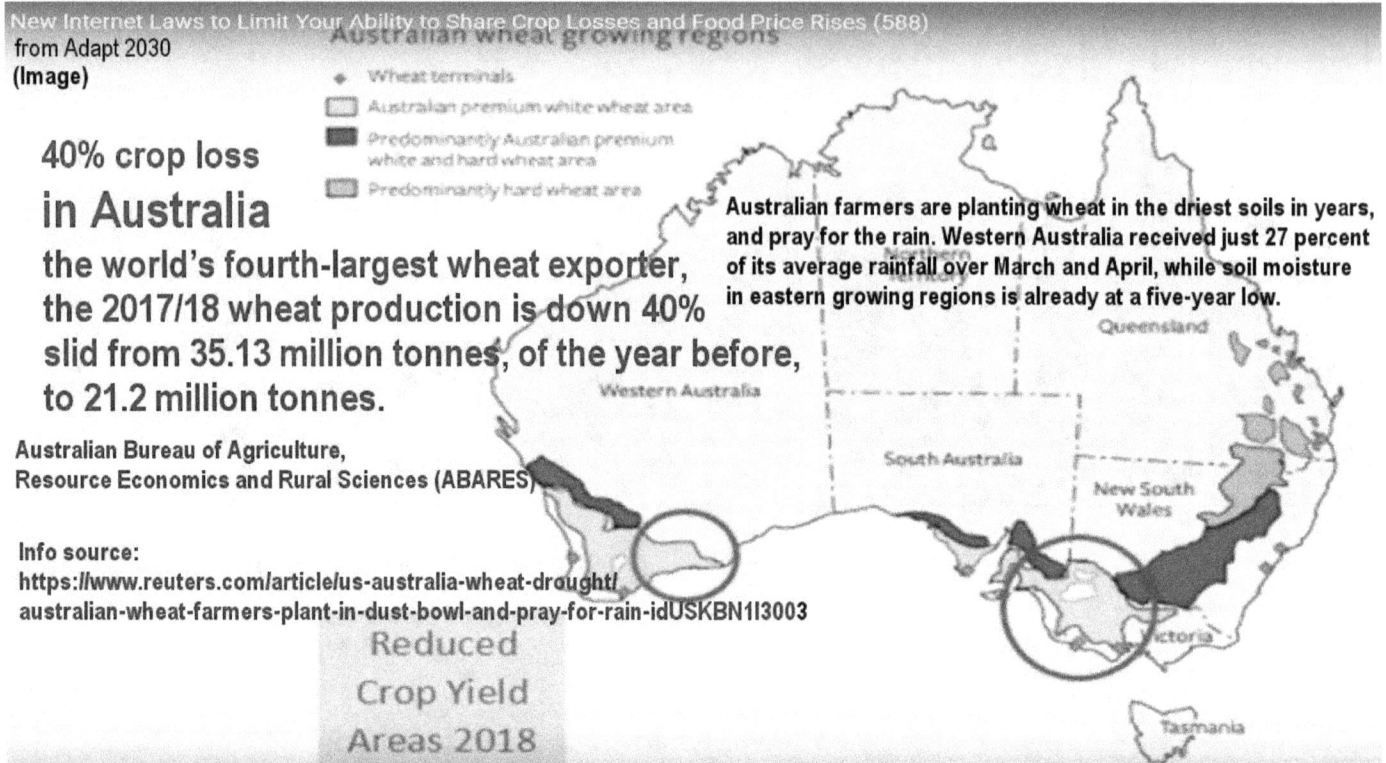

While the collapse is still only in the early stages, large-scale crop losses have already been experienced.

# Are we facing a large-scale population collapse then ?

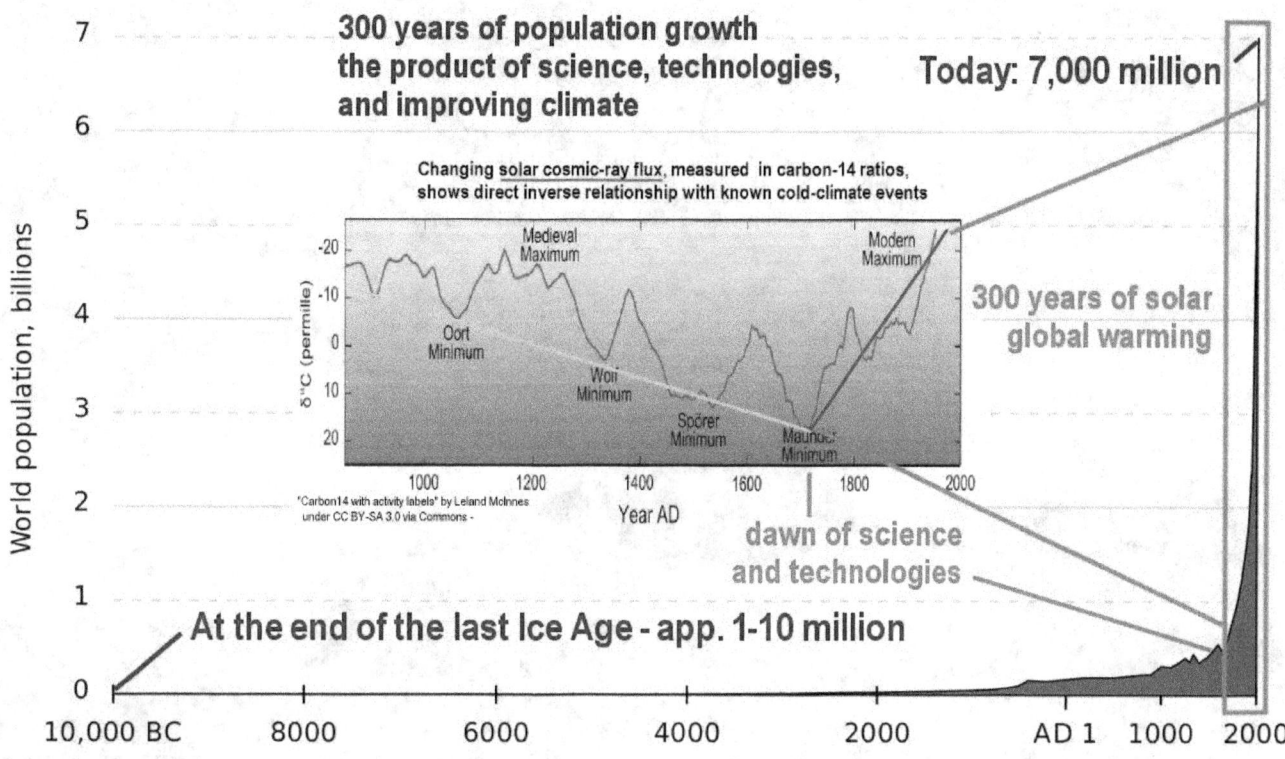

Are we facing a large-scale population collapse then, as the result of the climate collapse? Are we facing crop failures, and them fast increasing, in the years ahead?

I would answer no.

# We have the scientific and technological potential developed

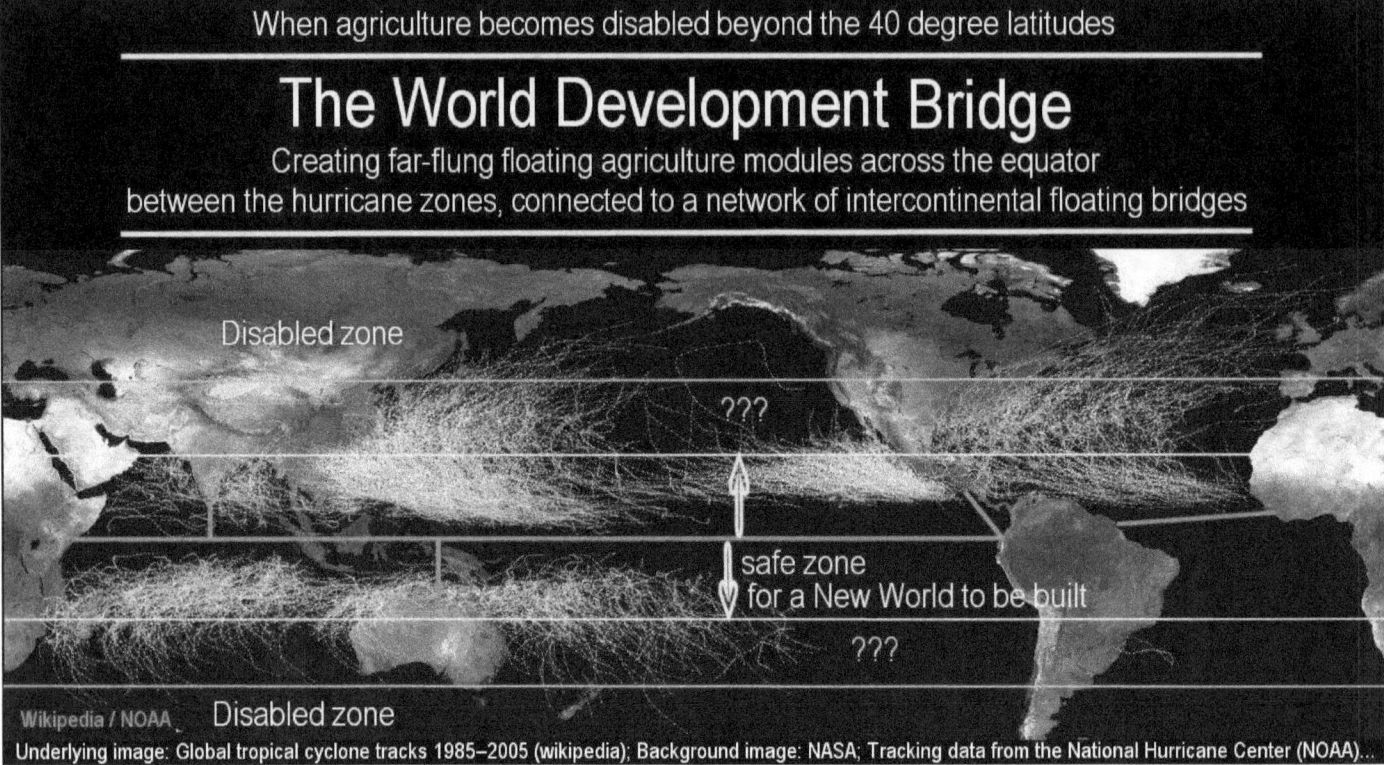

The answer has to be no, because we have the scientific and technological potential developed to lift agriculture out of its dependence on climate conditions. If it wasn't for this potential we would be seeing an enormous population collapse in the near future, such as we have never imagined.

But, as I said, I don't see this happening. This is so, because we have the power toady to avoid the population collapse that follows collapsing agriculture, by simply relocating our agriculture into the tropics and into indoor facilities there.

This power, that we have on this scene, is not trivial, and it is real. It is amazing and grand, and worth celebrating. When this creative power is applied, the human scene unfolds into joy, and peace erupts along the way.

There is no joy in being dead.

> **There is no joy in being dead.**
> **There is no peace in a world that is culturally dead.**
> **There is no power unfolding when science is dead.**

There is no joy in being dead.

There is no peace in a world that is culturally dead.

There is no power unfolding when science is dead.

# There is no real alternative to being alive with joy

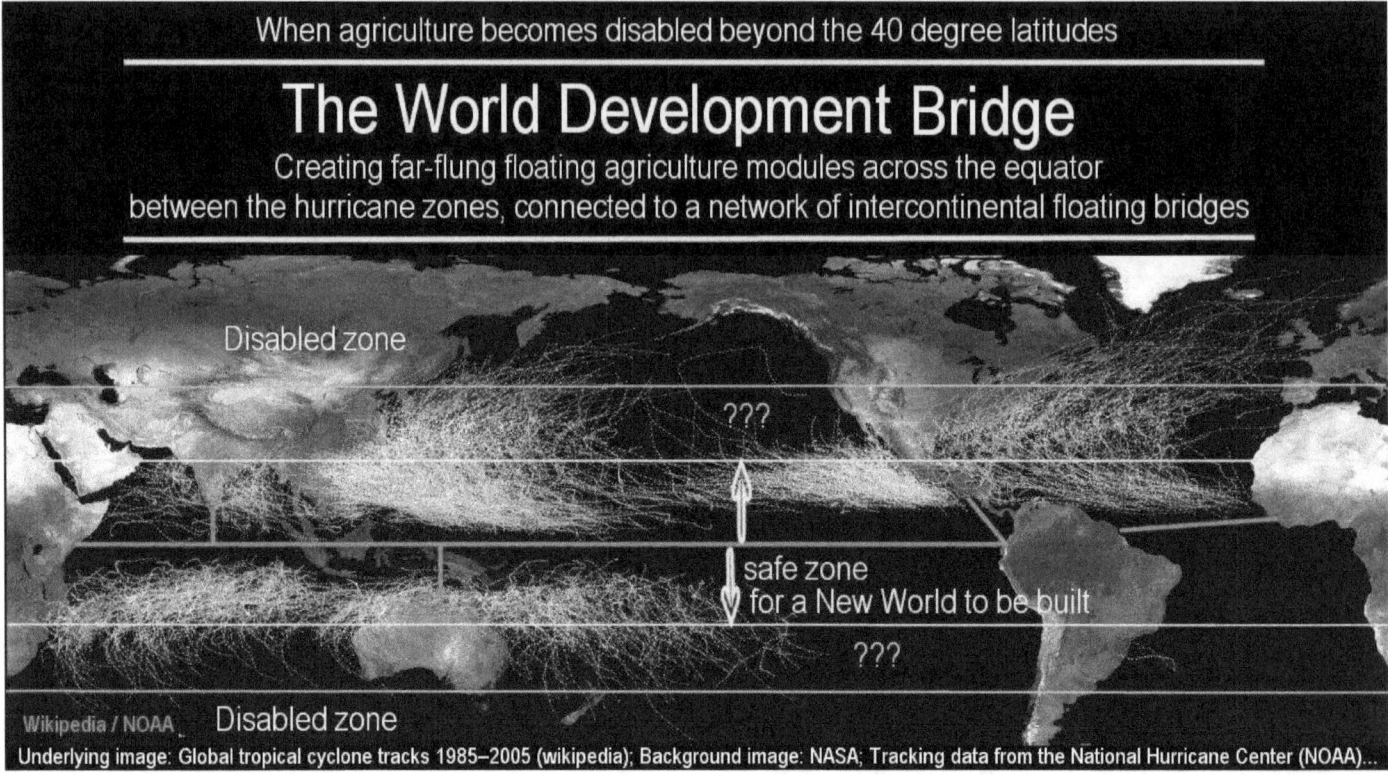

There is no real alternative to being alive with joy, and peace and power, because such an alternative is unthinkable and is unsurvivable.

# Only 1 of the 8 major human species has survived, and that's us

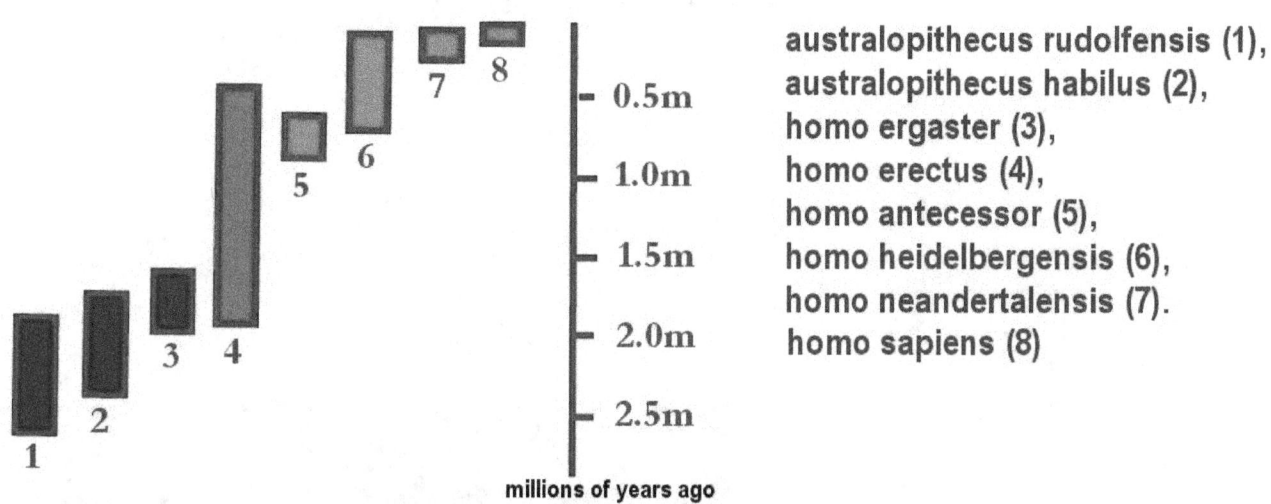

## Main human species

australopithecus rudolfensis (1),
australopithecus habilus (2),
homo ergaster (3),
homo erectus (4),
homo antecessor (5),
homo heidelbergensis (6),
homo neandertalensis (7).
homo sapiens (8)

millions of years ago

We, the homo sapiens (8), are the only surviving,
and the shortest lived of all the the human species,
at barely 200,000 years of age.

During the course of the harsh periods of the ice ages, or leading up to them, only 1 of the 8 different major human species that have developed over time, has survived, and that's us. Seven of the prior species have perished, most likely as the result of the many Ice Ages erupting along the way.

The longest-lived species was Homo erectus. But it too vanished. It likely perished during the period of the great population bottleneck that archeology tells us of, in which only a few hundred people had survived, worldwide, that we are all offspring of.

# Population Bottleneck roughly 150,000 years ago

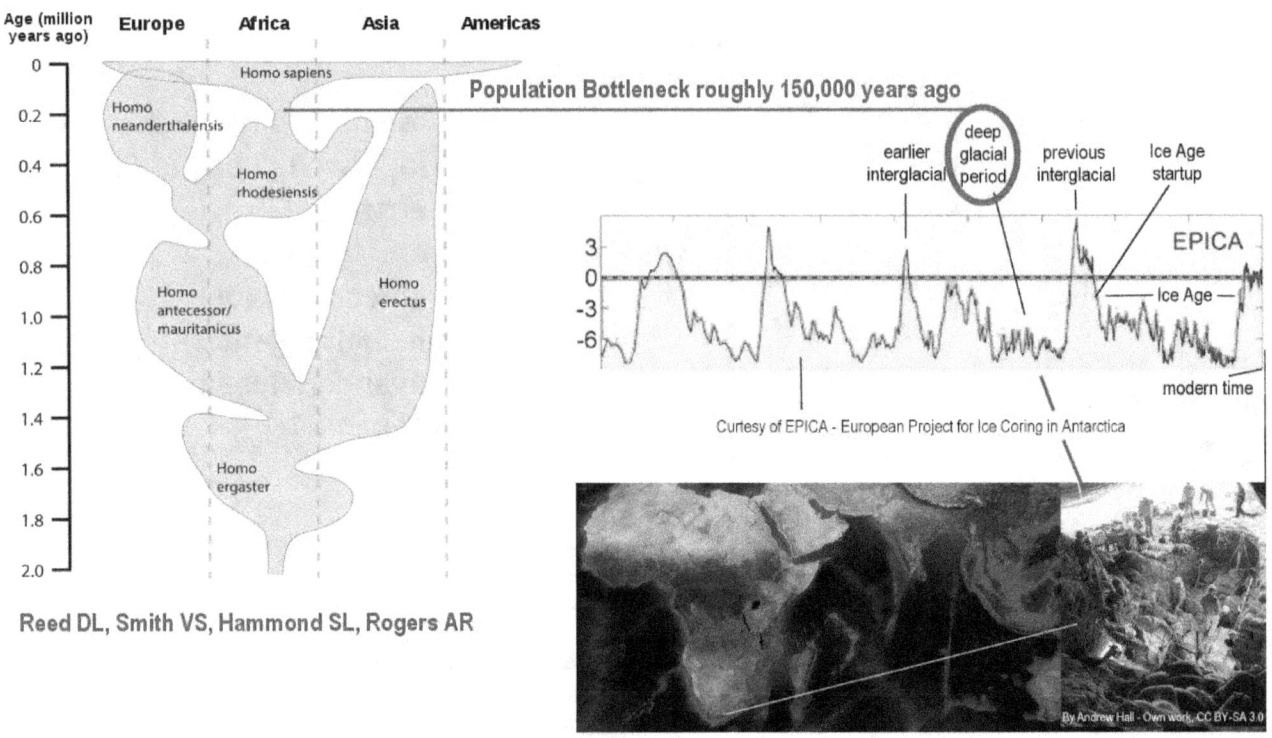

The Population Bottleneck occurred roughly 150,000 years ago.

Homo erectus may have perished during the deep cold glacial period of the second-last Ice Age, some time around 150,000 years ago. The climate was so harsh that humanity nearly ceased to exist. The critical timeframe became known in archeology as the period of the population bottleneck. In this period the human population collapsed from tens of thousands in number, to just a few hundred.

# In seaside caves at a place called Pinnacle Point

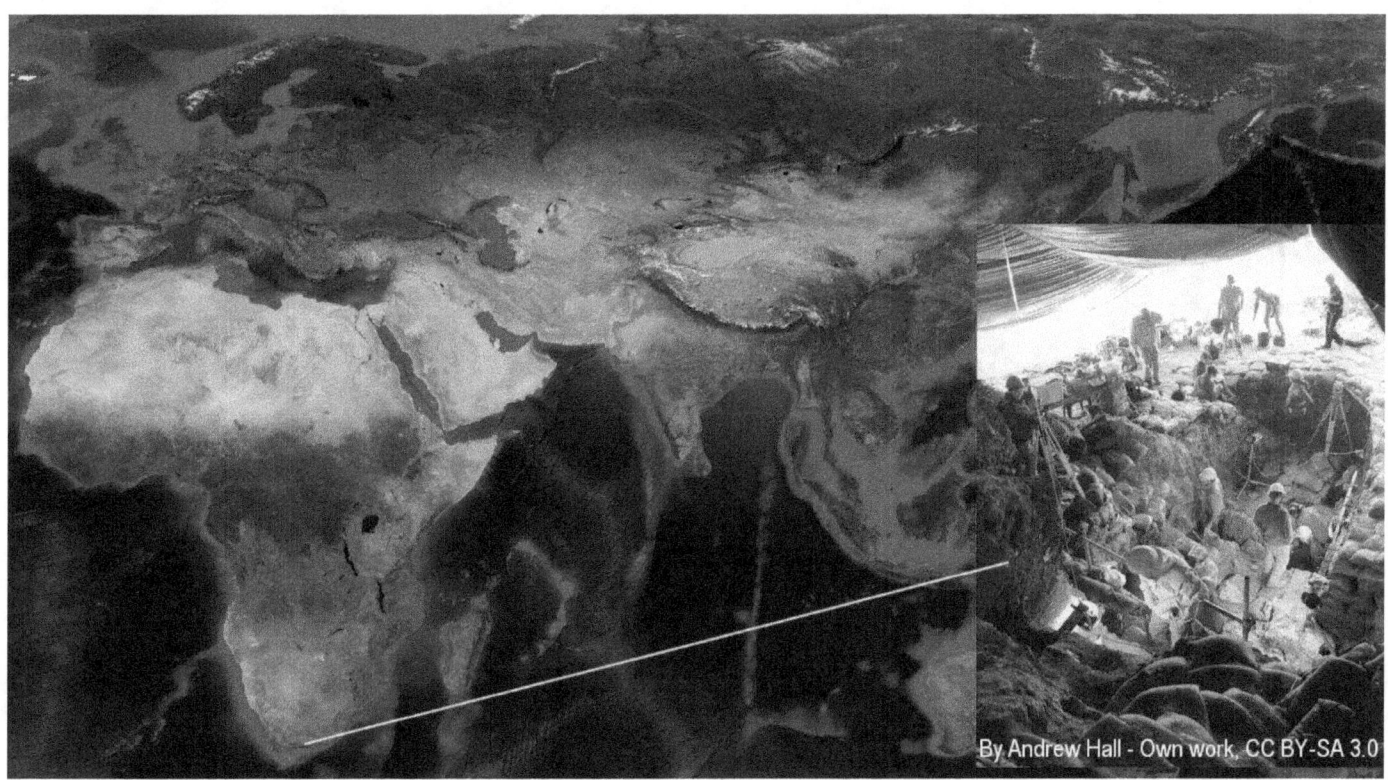

Those few were discovered to have lived in South Africa in seaside caves at a place called Pinnacle Point, where they survived being nourished by the sea.

# Devastating impact a major climate collapse can have

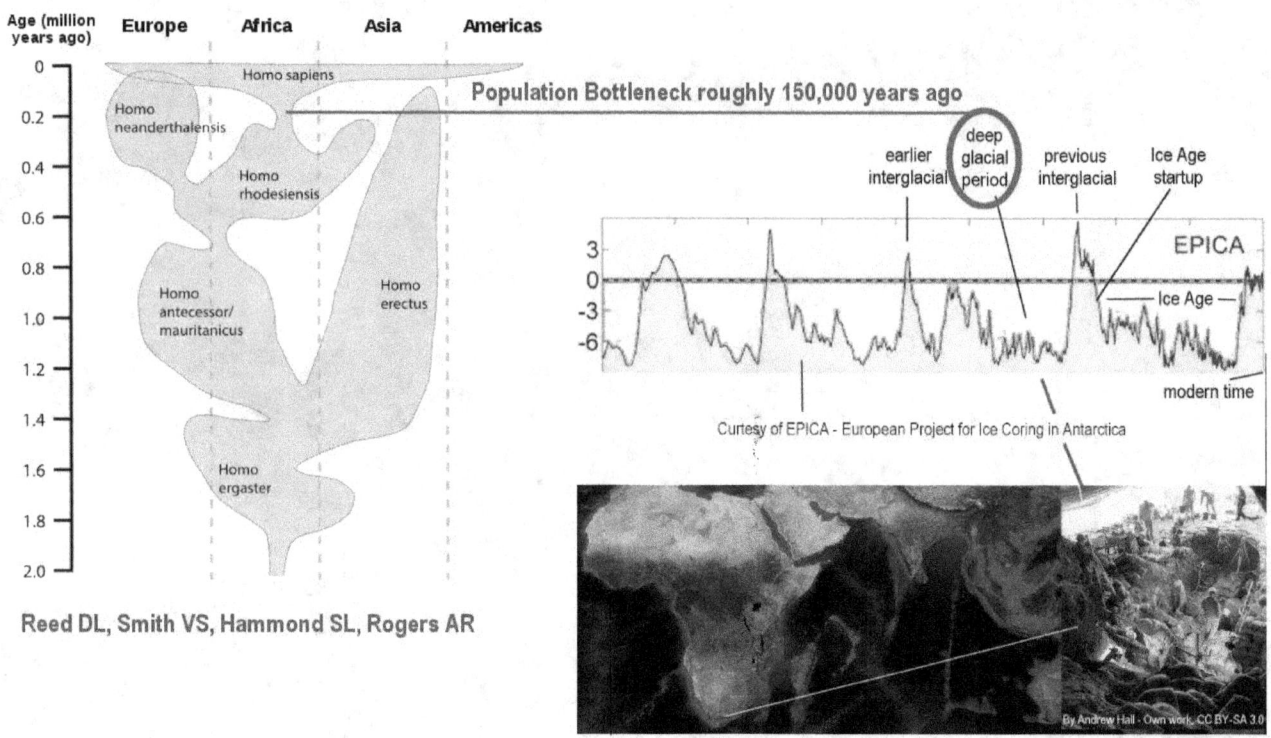

Reed DL, Smith VS, Hammond SL, Rogers AR

The population collapse during the bottleneck period illustrates rather dramatically the immensely devastating impact a major climate collapse can have, especially when it occurs during a deep glaciation portion of the Ice Ages.

While the survivors from the bottleneck recovered and rebuild their civilization, and then spread out across the world, the total world population that emerged into our interglacial at the end of the last Ice Age, around 12,000 years ago, was rather small in numbers, a minuscule 4 million people strong.

# This minuscule 4 million worldwide, was all that we had after 2.5 million years

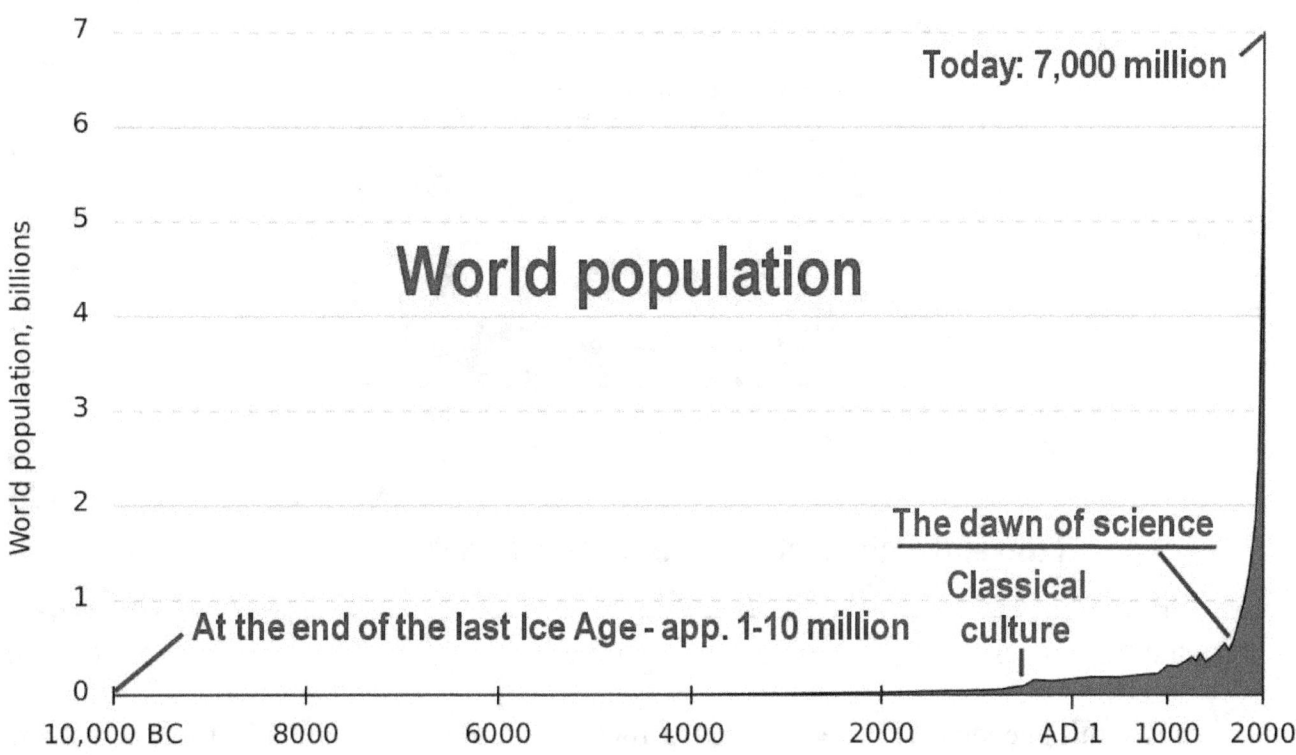

This minuscule 1 to 10 million world population, which modern research puts at 4 million worldwide, was all that we had to our credit after 2.5 million years of growing up. And even after the Ice Age had ended, when the climate recovered, the world population appears to have grown extremely slowly. The world population level remained essentially flat, until modern civilization began in the Greek Classical Period. And even then, it increased only slightly to half a billion people, between the Greek Classical Period and the Little Ice Age in the 1600s.

After that, all of a sudden, it doubled in only 2 centuries. We had a billion people living on our planet in the 1800s. Then, bang, another 2 centuries later, the world population had risen from 1 billion to 7 billion.

# What had caused this increase from the 1800s onward

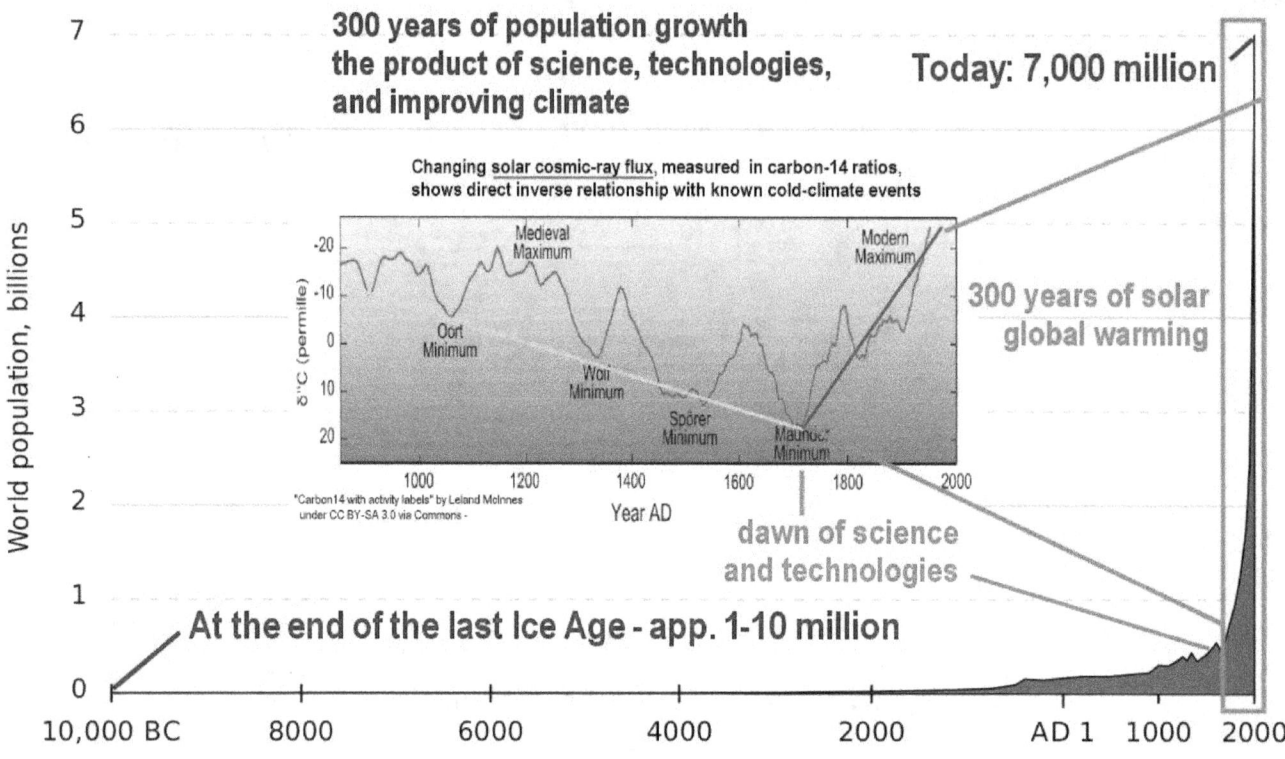

What had caused this phenomenal increase from the 1800s onward, was evidently the triple combination of factors that I have mentioned before - the combination of science, industrialization, and the massive global warming by the Sun that occurred in the same timeframe. These three factors, all flowing together, had increased the potential population density of our world. The realization of this potential, of course had enabled the phenomenal population increase that has occurred. But one of the three factors is now beginning to fail.

The factors of energy intensive mechanization in farming and related scientific advances are still continuing. These factors are manmade. They are under our control. However the third factor is not. The global-warming climate recovery from the 1700s onward, has a cosmic cause that affects the Sun. This, we cannot control.

When the three factors came together as they did almost simultaneously, the result was ample food had enabled the industrial revolution, which in turn increased farm productivity even more.

The Sun, getting stronger in solar activity, had enabled a renaissance in living

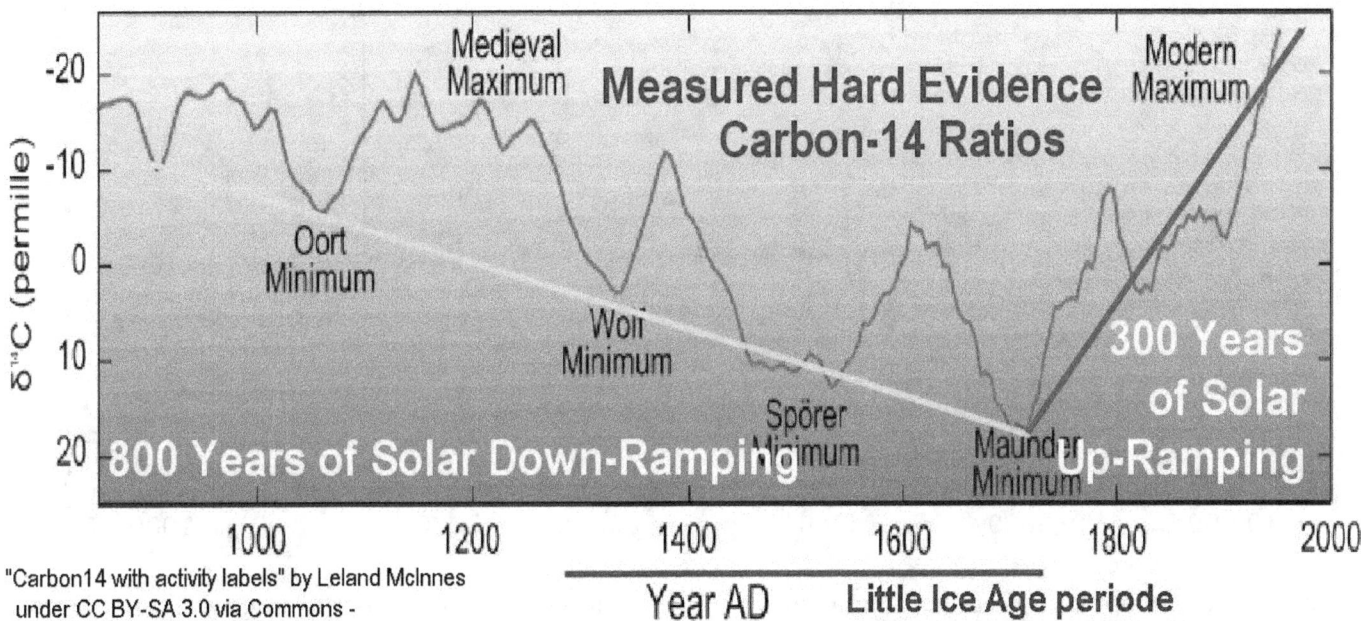

But it all happened against the background of the global-warming climate recovery by the Sun. Without that solar recovery, none of what has been achieved would have been possible. The Sun, getting stronger in solar activity, had enabled a renaissance in living on Earth, both in physical abundance and in optimism for the future, and all this, because the Sun had enabled the manmade factors to become amazingly effective. It brought joy to the world.

# The recovery of the climate had opened up the portal for science

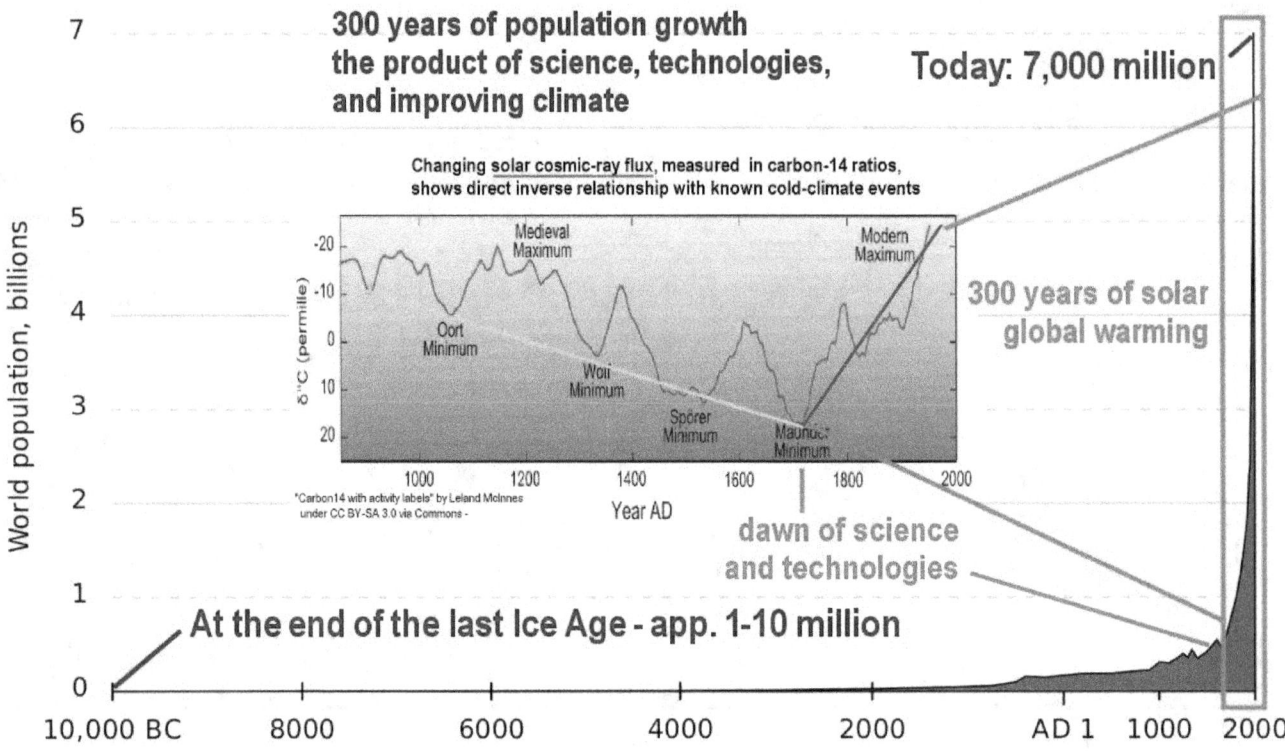

The phenomenal population increase that resulted in the period of the solar up-ramping, would not have been possible without the global warming by the Sun. The recovery of the climate had opened up the portal for science and technology to become effective, that were all taking off at the time.

# Now that the climate recovery by the Sun is over

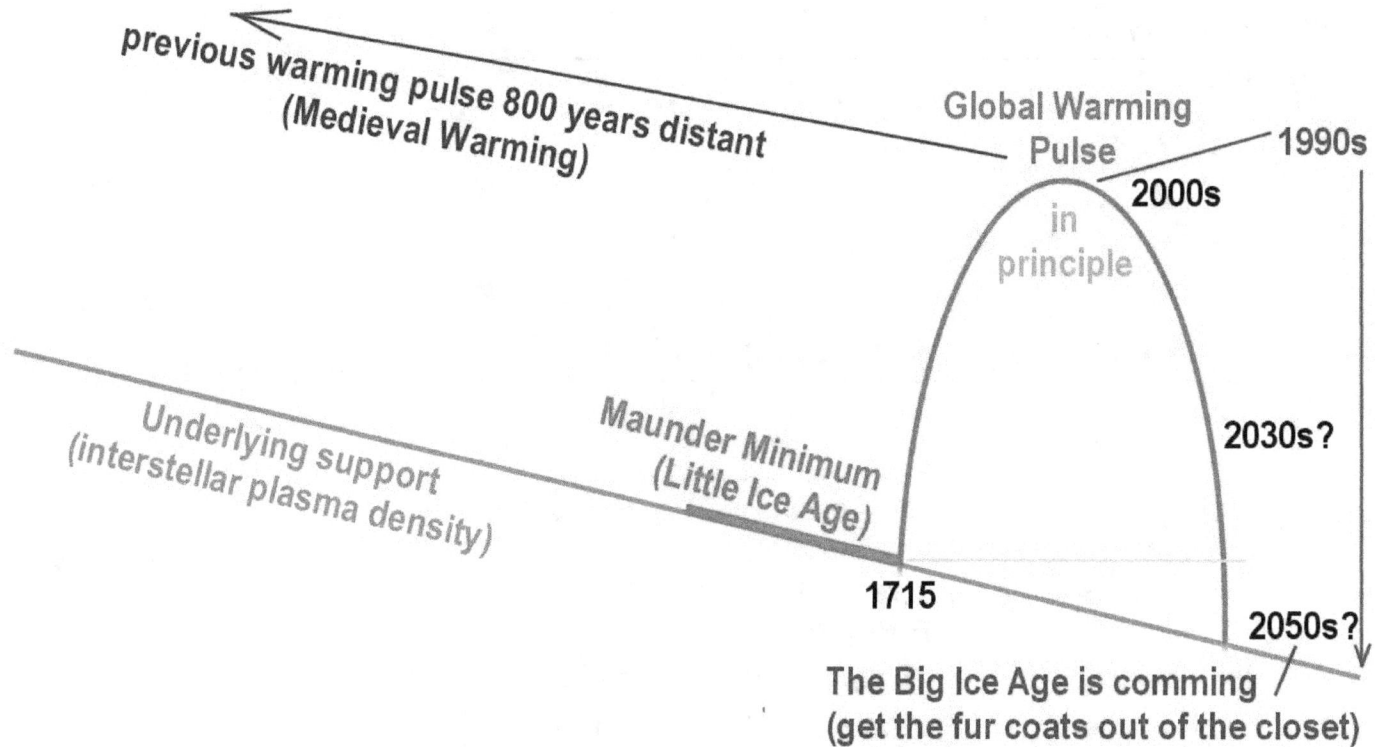

Now that the climate recovery by the Sun is over, we are on the fast track of the Sun dropping back to the level of the Little Ice Age in the 1600s that the global warming started from. A global climate collapse has now begun. It began faintly at first at the end of 1990s, but is now fast progressing, and I mean, fast.

# Rate of collapse reflected in the collapsing solar-wind pressure

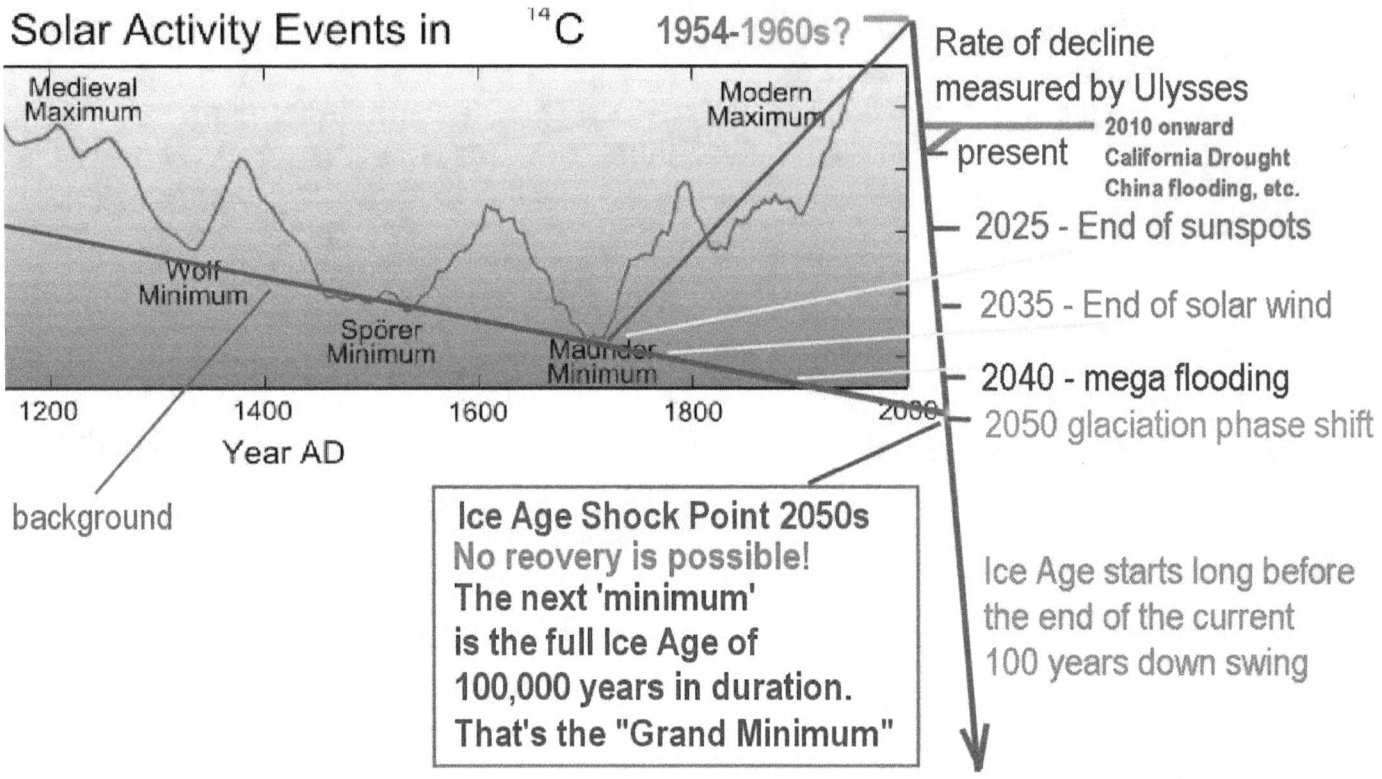

The rate of collapse of the solar activity, is reflected in the collapsing solar-wind pressure that the Ulysses spacecraft has measured.

# The spacecraft measured a 30% collapse per decade

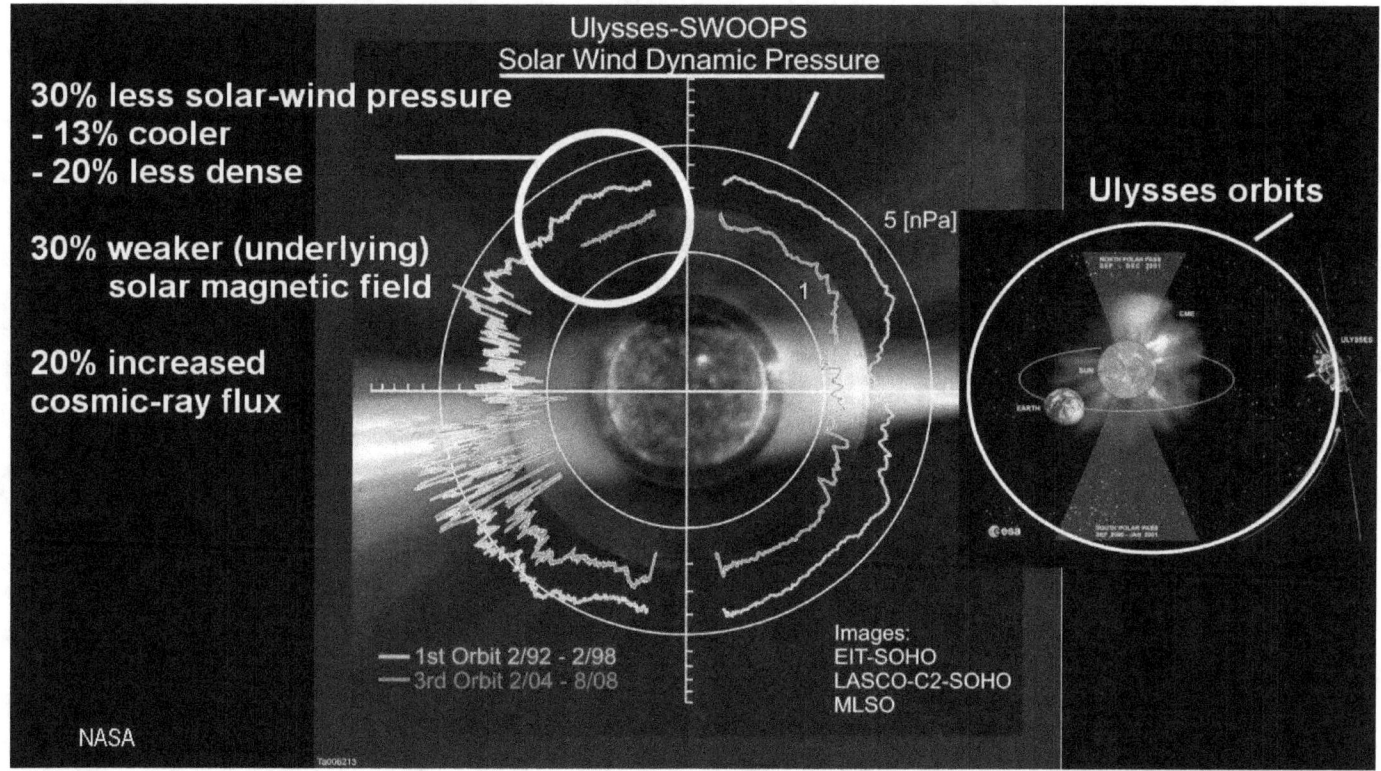

The spacecraft measured a 30% collapse of the solar wind pressure per decade, while it orbited the Sun. The collapse of the solar wind pressure reflects the collapsing solar activity. The result is a colder Earth.

# The solar-activity collapse tells us

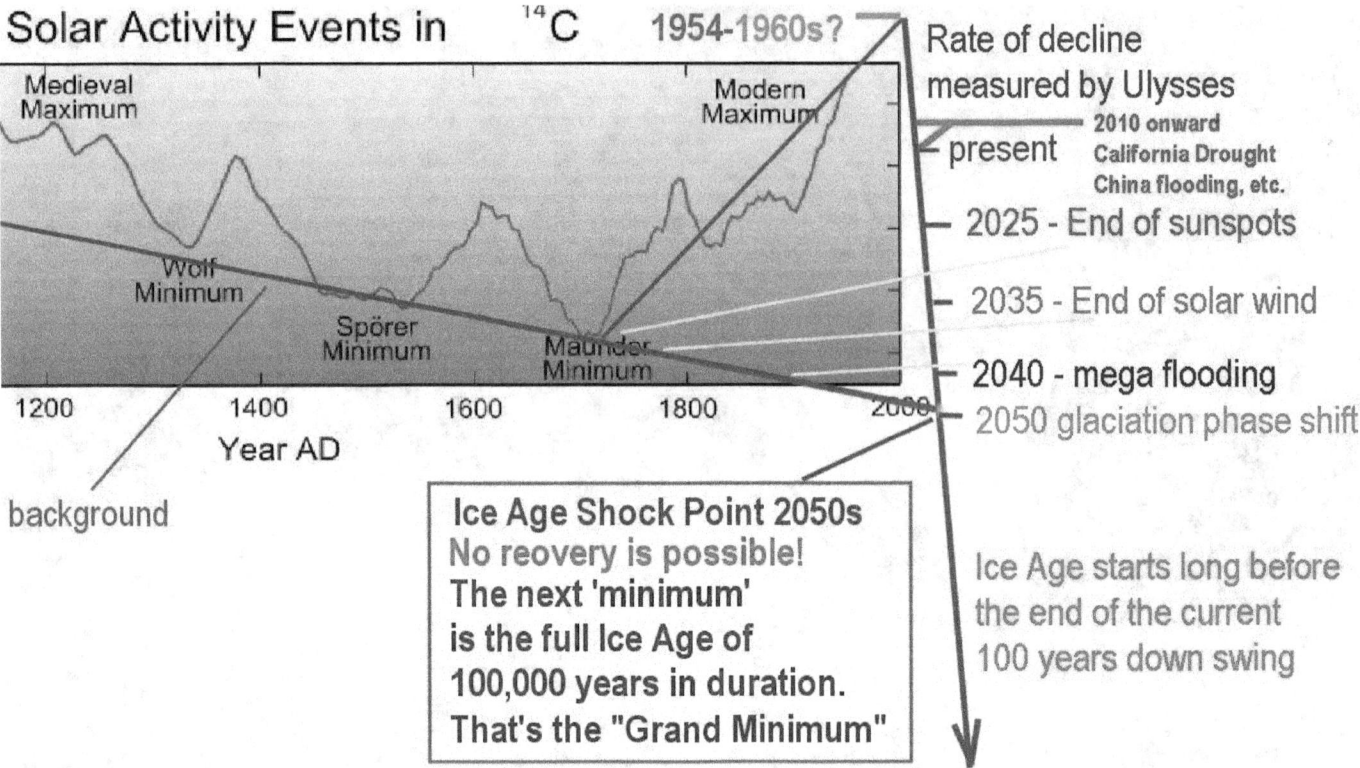

The solar-activity collapse tells us that the entire global warming gain by the Sun, that had occurred during the last 300 years, will likely be reversed by the time we get to the 2030s or 2040s. This means that the now ongoing climate collapse may be happening up to 10-times faster than the solar up-ramping had been progressing during the last almost 300 years.

# Climate collapse will affect agriculture in a big way

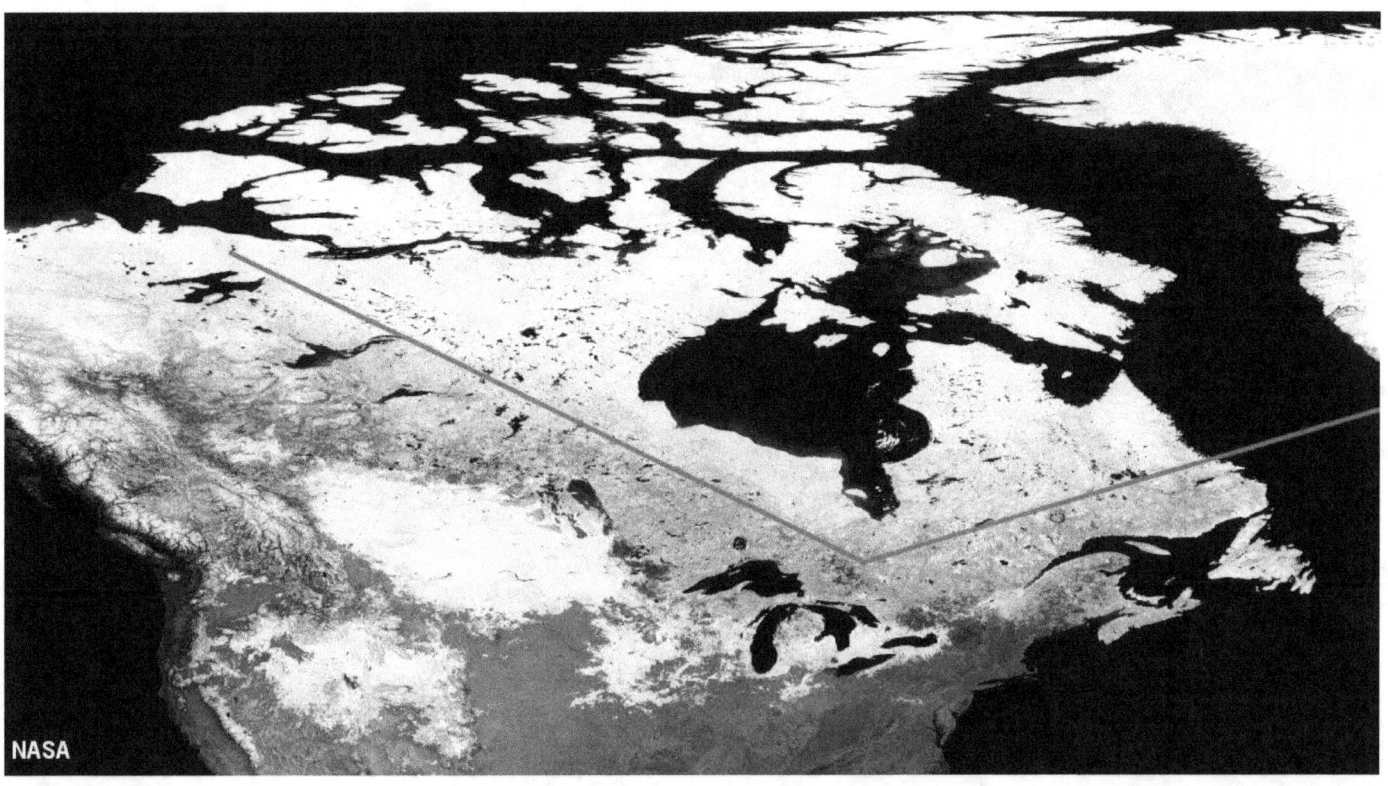

Of course, the fast climate collapse will affect agriculture in a big way, especially when the large cold air mass in the Arctic gets ever-colder, so that its outflow to the South begins to devastate the North American grain-growing regions.

# April 2018 winter-type blizzard, named Xanto,

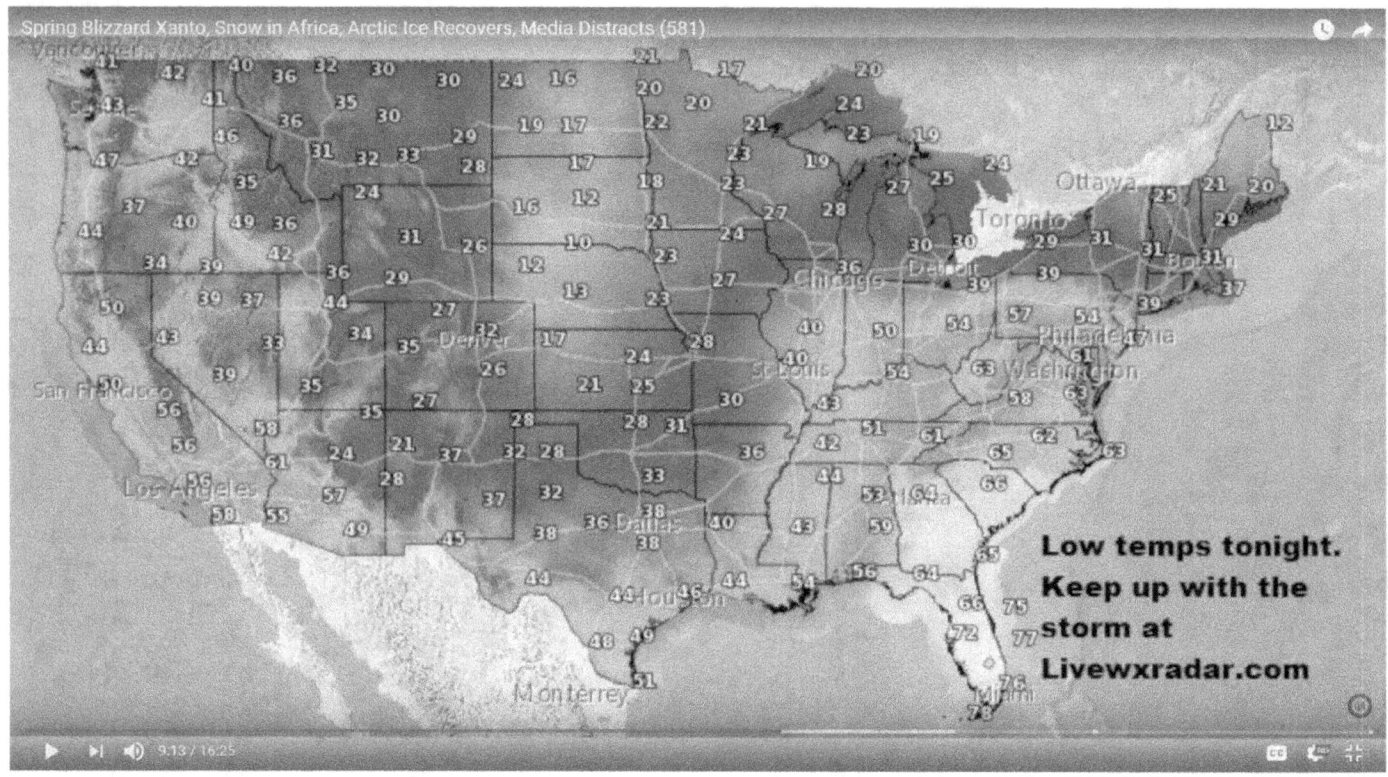

We saw the beginning of this already in mid-April 2018, when a winter-type blizzard, named Xanto, dumped snow and cold across the grain belt at a time when spring planting should have been in progress.

# The natural growing window is shrinking

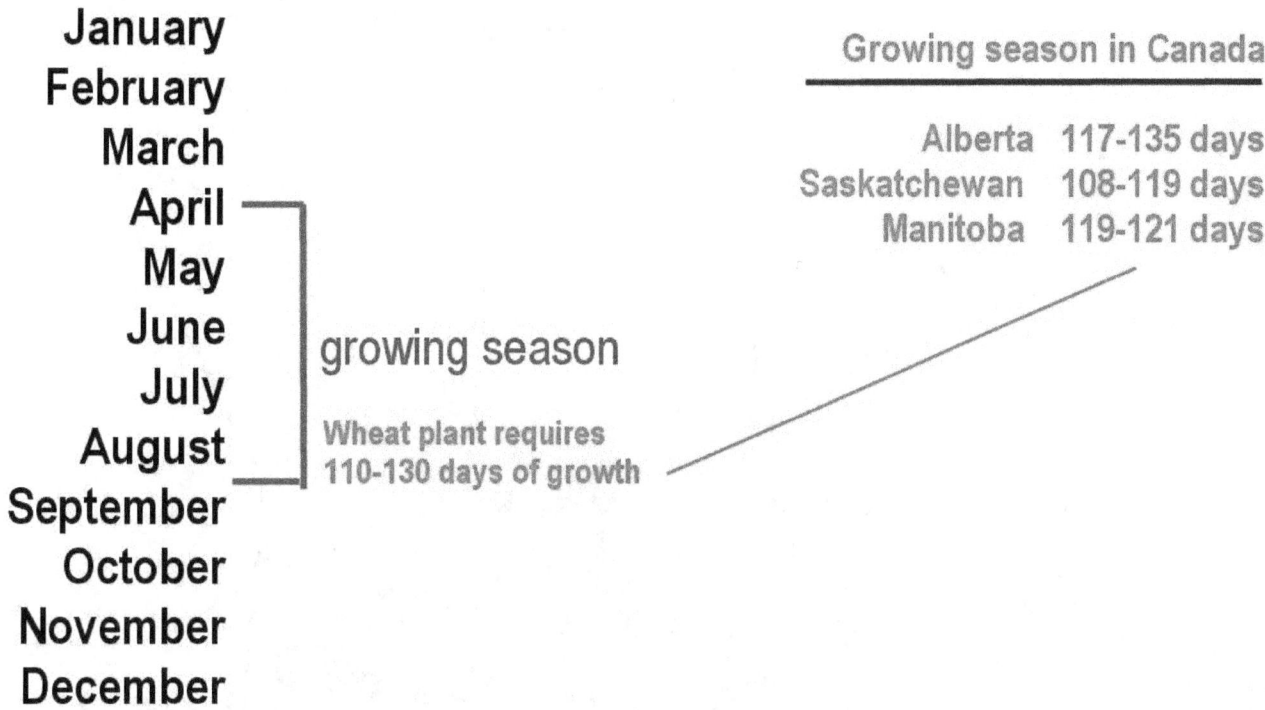

It appears that agriculture survived the late blizzard, this time. The remaining months of May, June, July, and August into mid-September, provide a long-enough growing season for wheat and corn to grow and mature. Corn requires 60 to 100 days to mature, and wheat 110 to 130 days. At the moment the natural growing window is wide enough in most places to accommodate this requirement. But the window is shrinking.

# Climate collapse will shrink the growing season

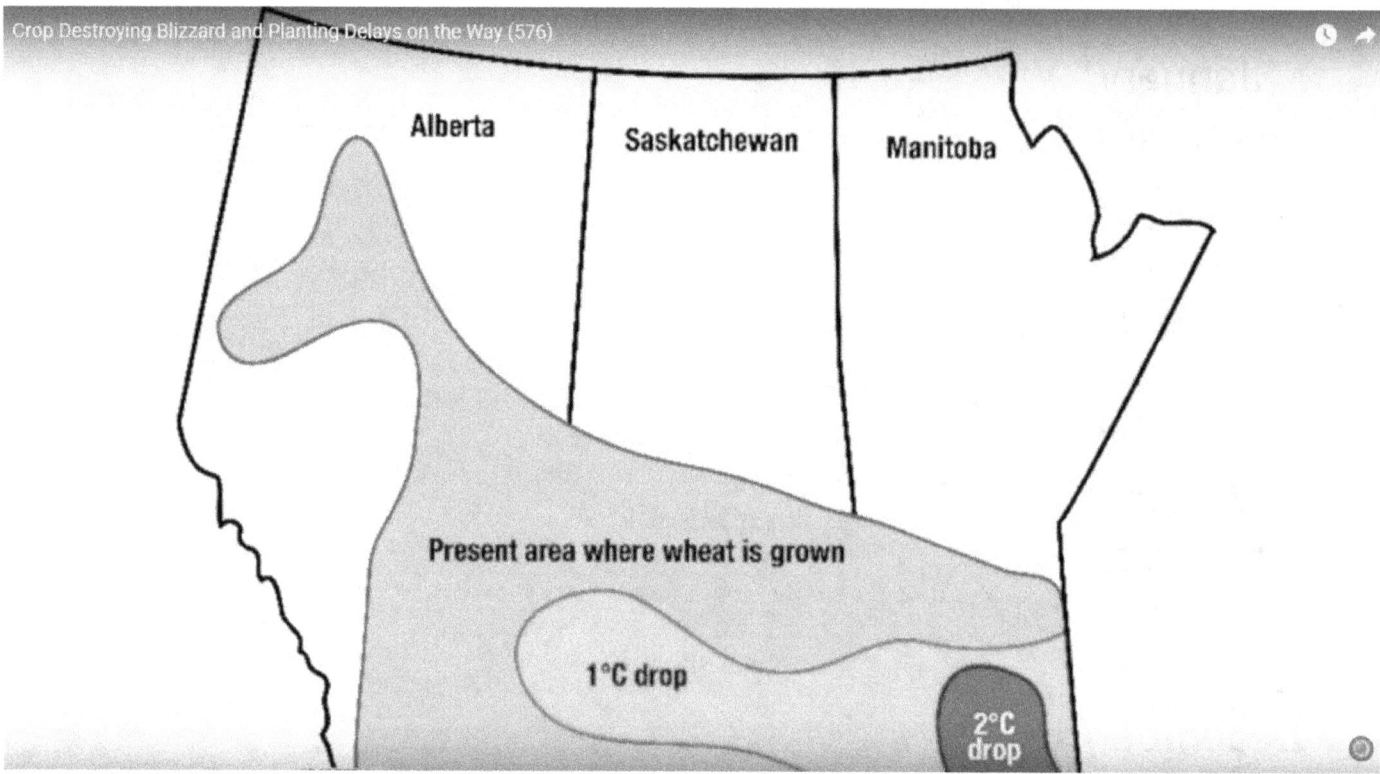

In the study shown here, for Canada, the area where wheat can be grown will shrink dramatically as the climate collapse advances and takes its toll.

As the climate collapse is advancing, it will shrink the growing season. The study suggests that the green area will be the first to become disabled for agriculture. We may get to this point in 5 years time. The blue area may become disabled in 10 years, and the rest in 15 years, approximately. While the precise timing is uncertain, the collapse will happen. The movement towards it is already in progress.

# Nor will the climate collapse affect only Canada

Nor will the climate collapse affect only Canada. The devastating cold will affect almost the entire grain growing region, extending deep into the USA, just as the blizzard Xanto had in April 2018.

# Europe is similarly affected

Europe is similarly affected. Its cold-flow comes from the East. They call it the Beast of the East. Major crop failures have already been experienced.

# The collapse in solar activity is continuing

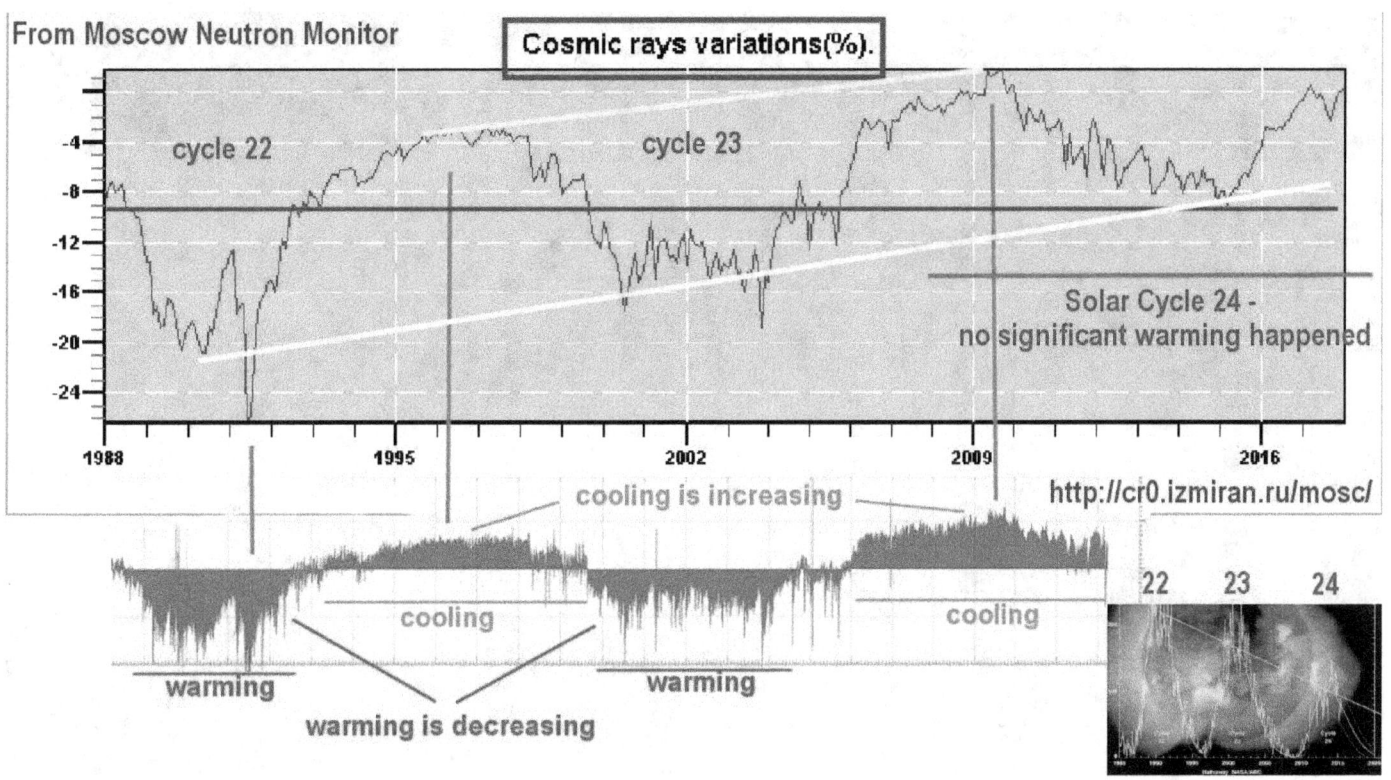

And as I said, the end of it all promises to be worse. The collapse in solar activity that the Ulysses spacecraft had measured, is continuing at the same rate. This has been measured by Moscow Neutron Monitor, which measures the neutron density in the atmosphere that is produced by solar cosmic-ray flux, which in turn reflects solar activity intensity.

And, as the solar cosmic-ray flux goes up, the earth gets colder,

# Increasing volumes of solar-cosmic ray flux cause increased cloudiness

Increasing volumes of solar-cosmic ray flux cause increased cloudiness. The white top of clouds reflects larger portions of the incoming sunlight back indo space, which cools the Earth.

# Increased cloud nucleation also causes faster rainout

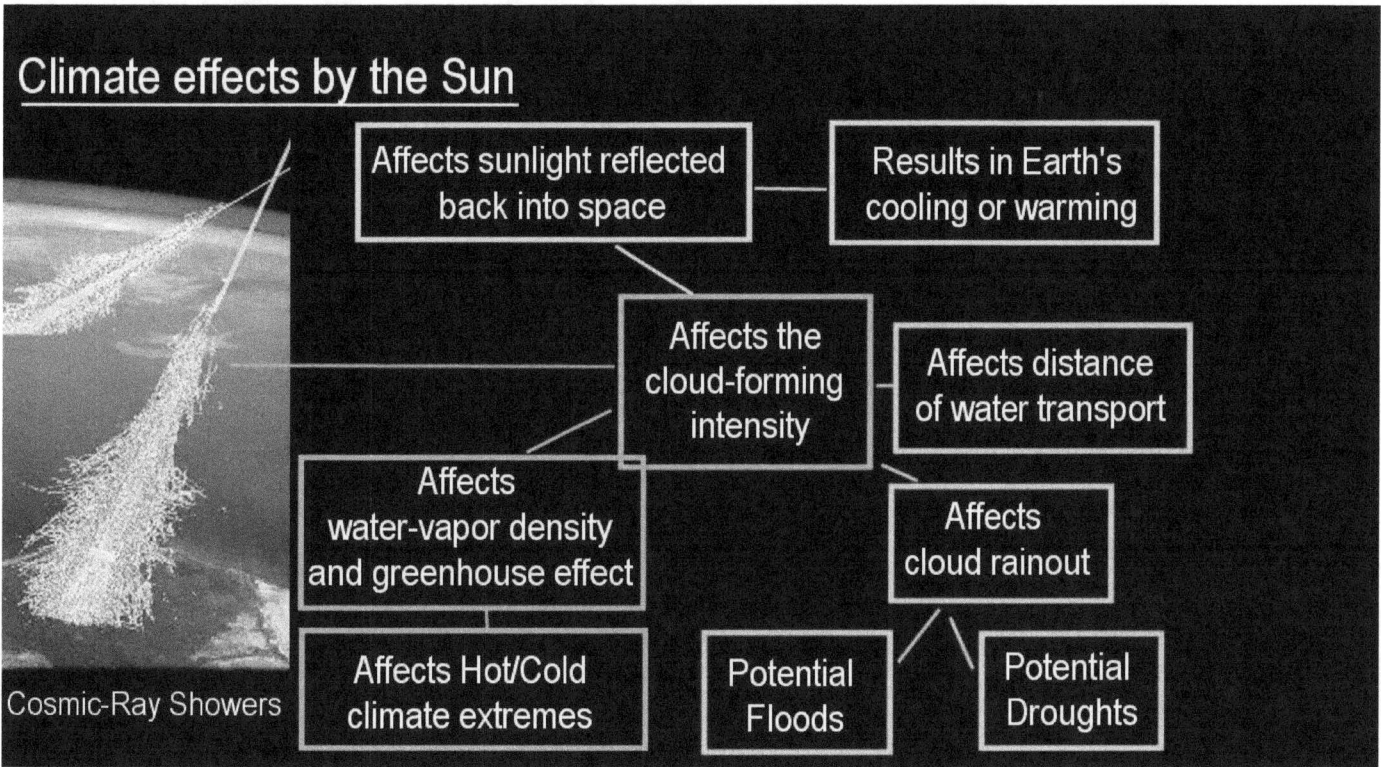

Increased cloud nucleation also causes faster rainout of the clouds. Then, with less water vapor in the air, droughts begin to develop. And that's what we see happening. Droughts are increasing, and occurring in evermore places.

# Even the greenhouse effect is diminishing

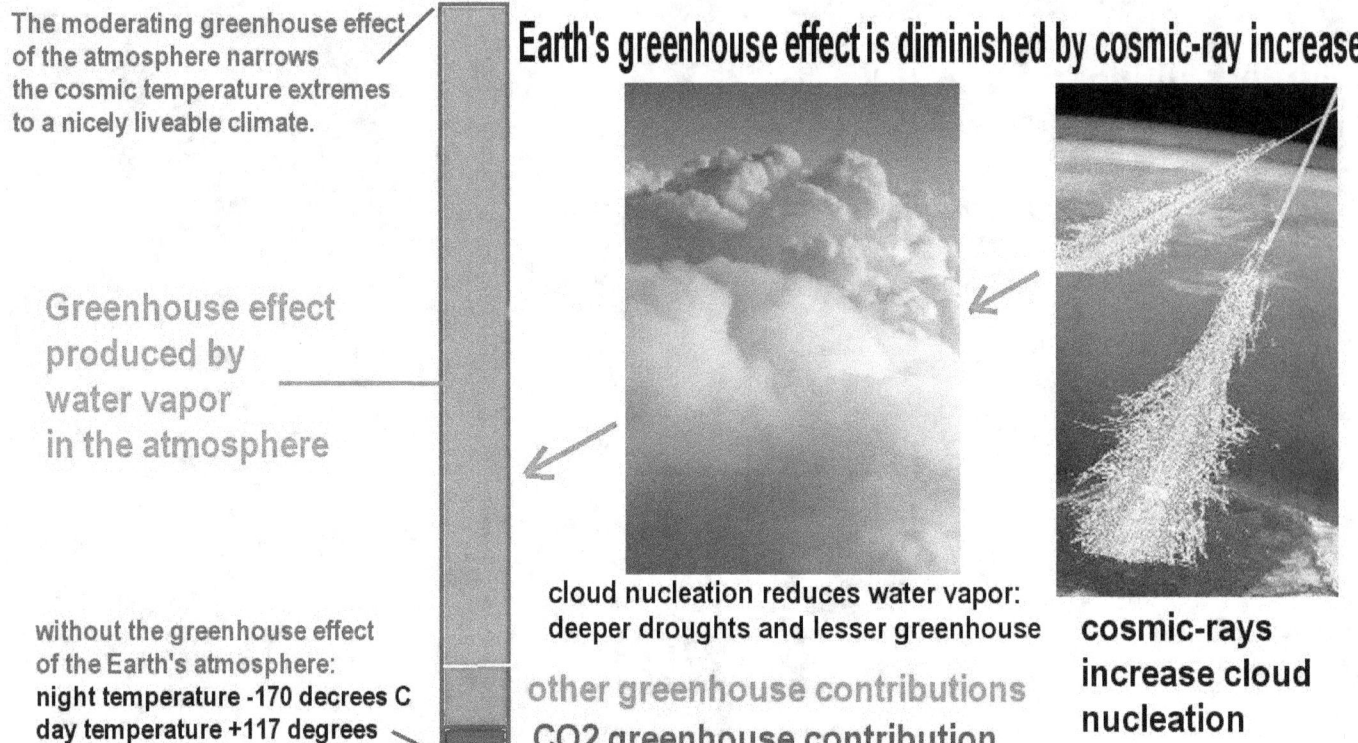

Even the greenhouse effect is diminishing, which is largely caused by water vapor.

# Climate is now collapsing back to the Little Ice Age

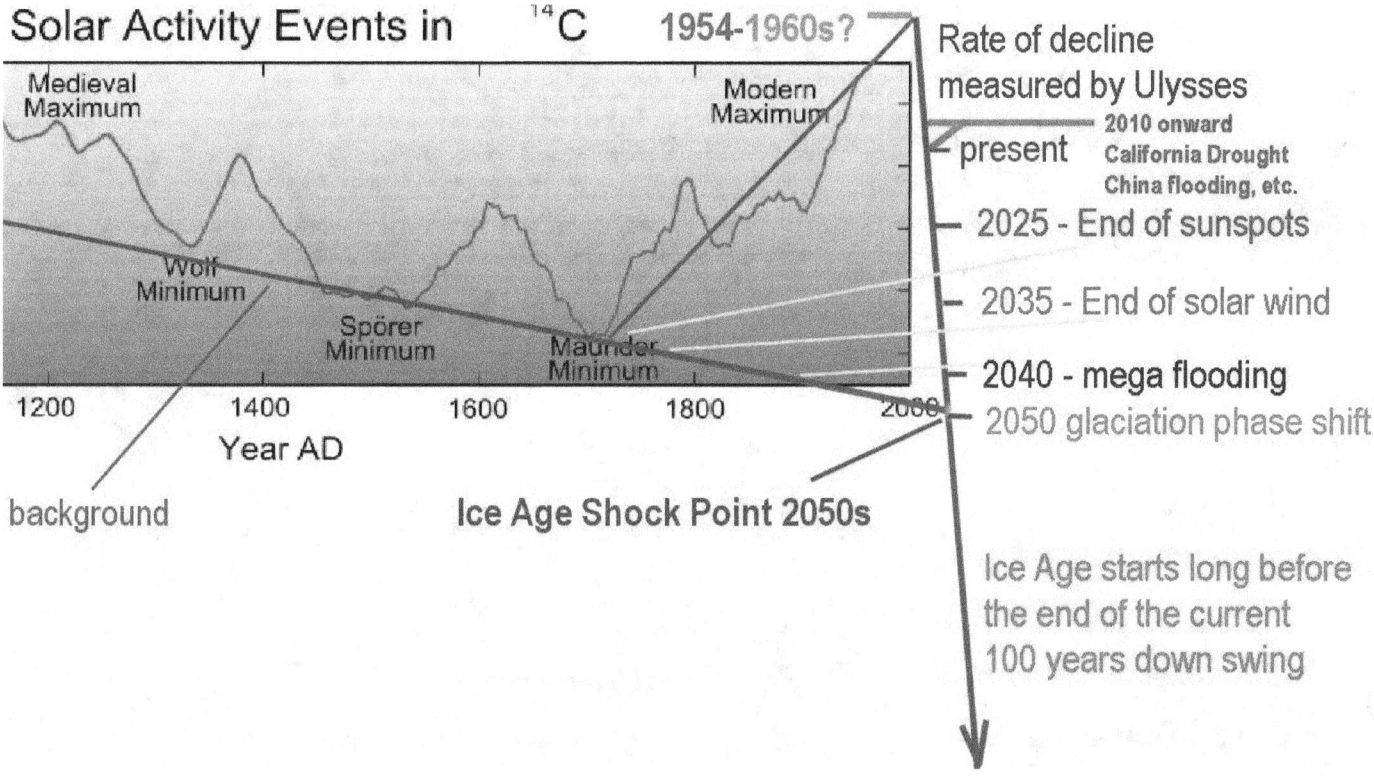

The climate is now collapsing back to the level of the Little Ice Age of the time of the Maunder Minimum, which gave us the coldest climate on record in the last thousand years, with the densest cosmic-ray flux occurring, and as a consequence, extreme drought conditions.

When the climate collapses back to the conditions of the Little Ice Age, we may see a corresponding population collapse happening. In the 1600s, the world population was small, at roughly half a billion people. But this time the climate collapse won't stop there. It will fall below the level of the Maunder Minimum, all the way into glaciation. The resulting population collapse will then be unimaginable, if we allow it to come to that.

# We cannot escape or halt the now ongoing climate collapse

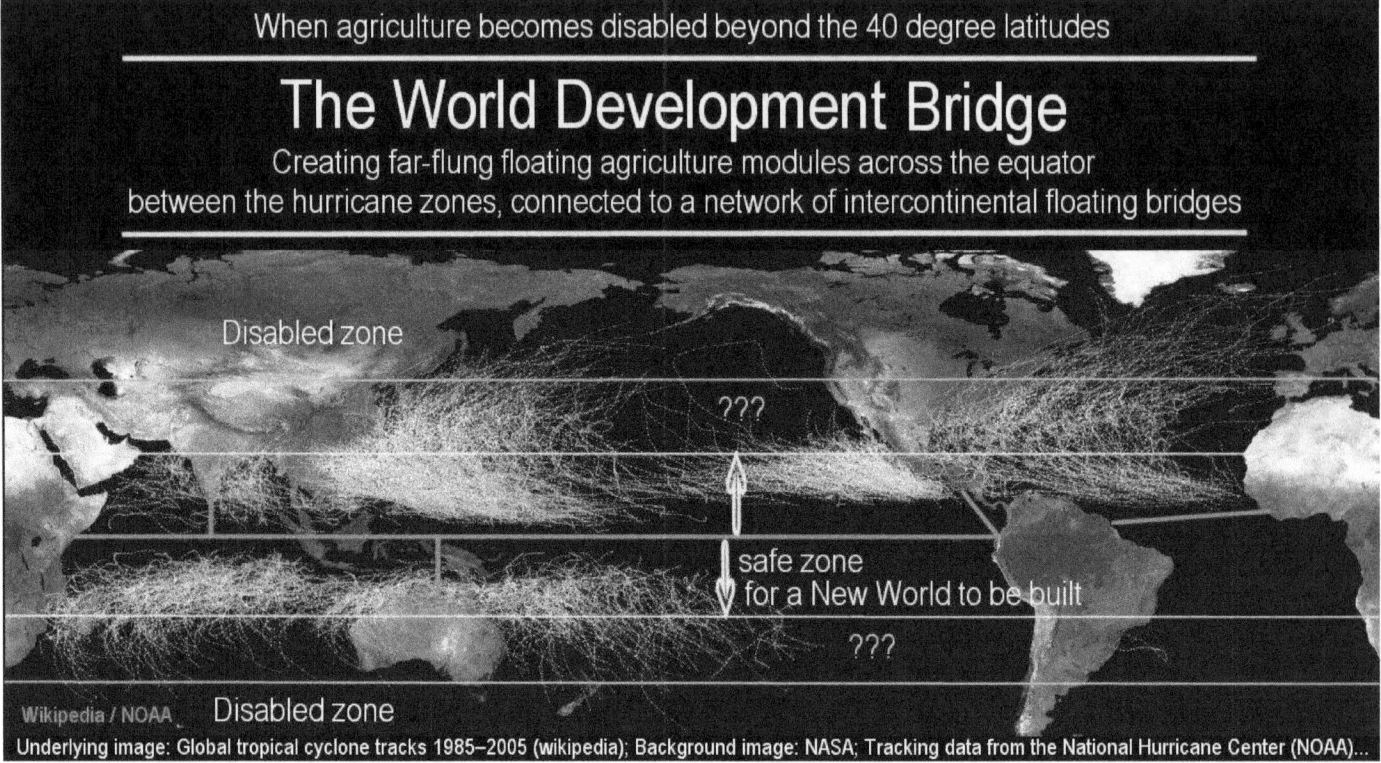

Fortunately, while we cannot escape or halt the now ongoing climate collapse, which is caused by dynamics that affect our Sun, we do have the technological capability today to avoid the consequences of the climate collapse by building us a New World with technological infrastructures that the climate collapse cannot affect.

This power, that we have, to build us a new world for continued living, should inspire tremendous joy. The liberation that we can thereby achieve is comparable to a prisoner on death row being released to life.

# When we drop below the Little Ice Age level, into glaciation

When we drop below the Little Ice Age level, into glaciation, we have to make do with 70% less solar radiation.

# We also have to make do with 80% less rain

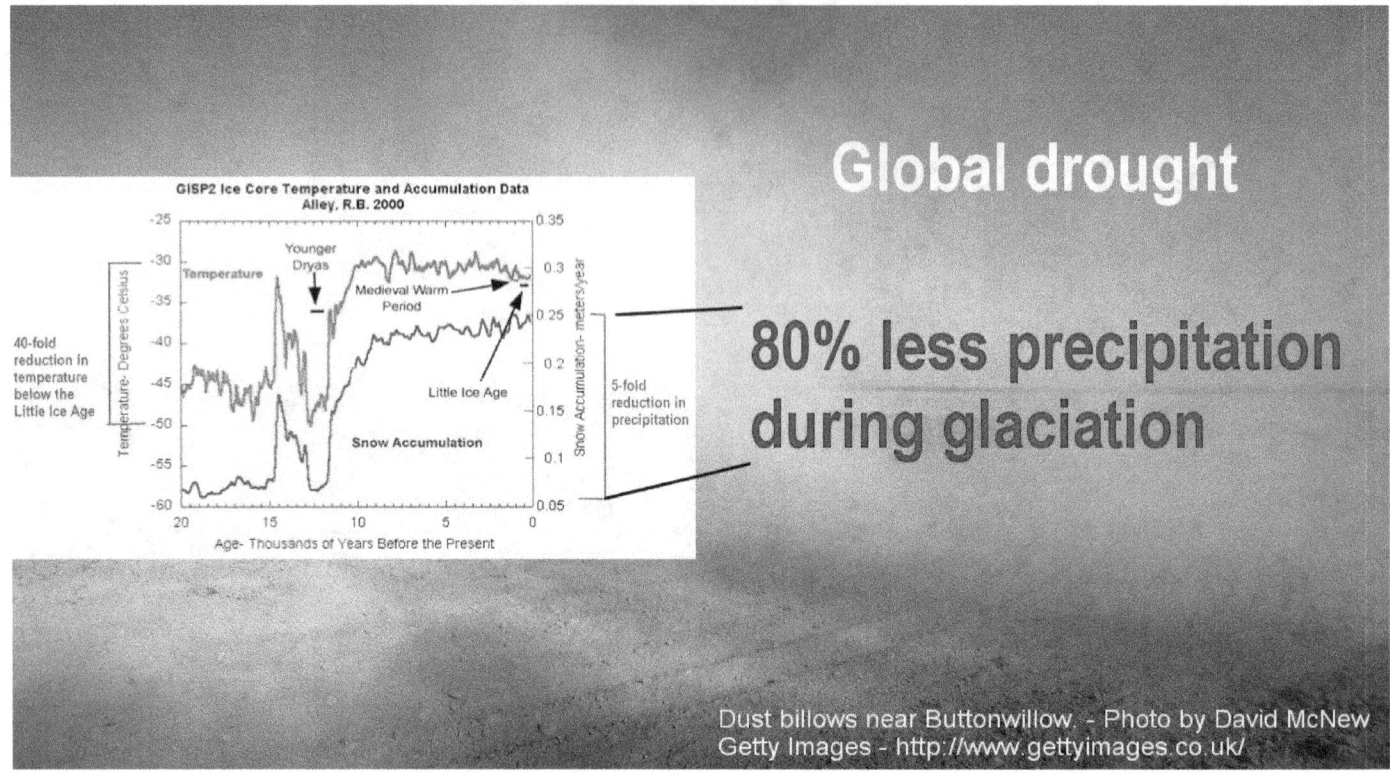

We also have to make do with 80% less rain, according to what the ice-core records tell us.

# To build us a new world synonymous of escaping a death sentence

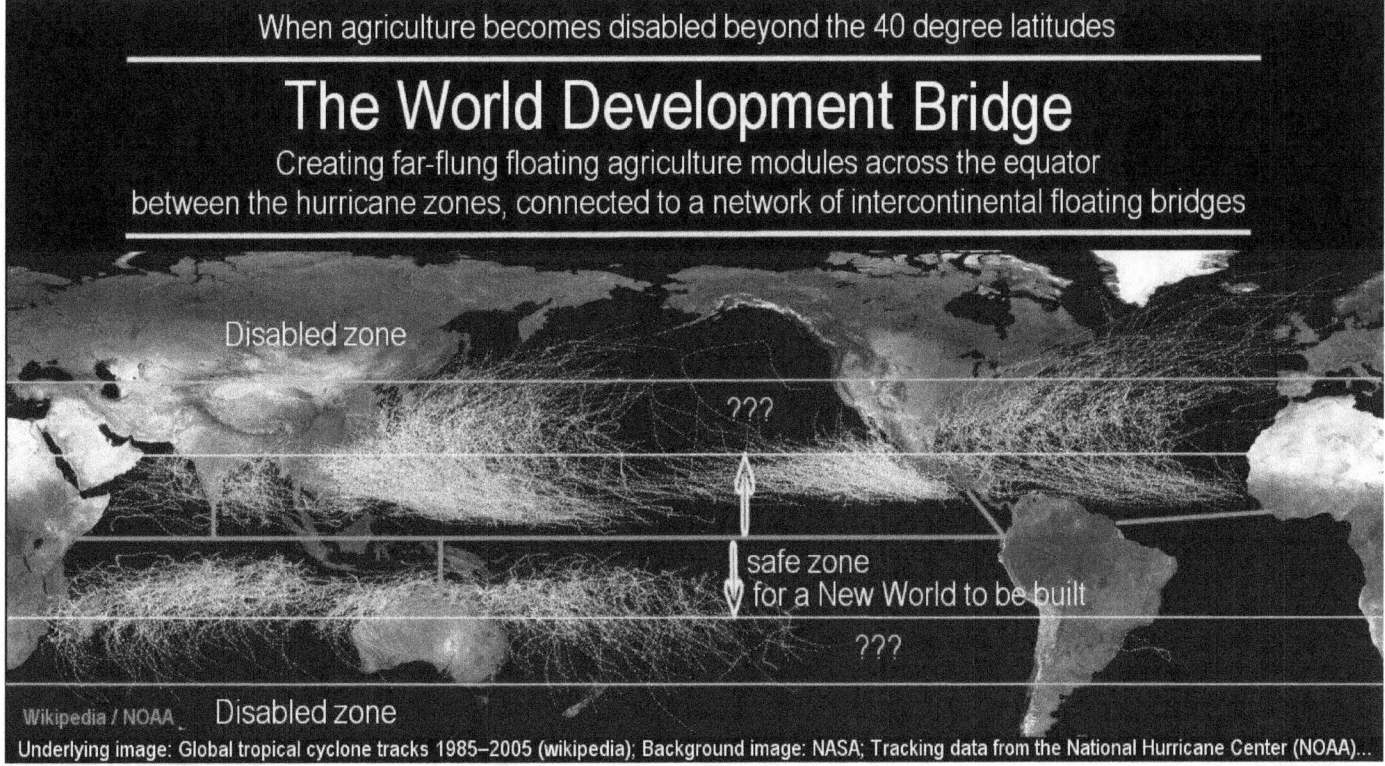

The final climate collapse of this magnitude would put the whole of humanity onto death row, if it wasn't for the power that we now have, to build us a new world where the change in climate is compensated technologically. That's synonymous of escaping a death sentence.

# We didn't have this power developed in the 1600s

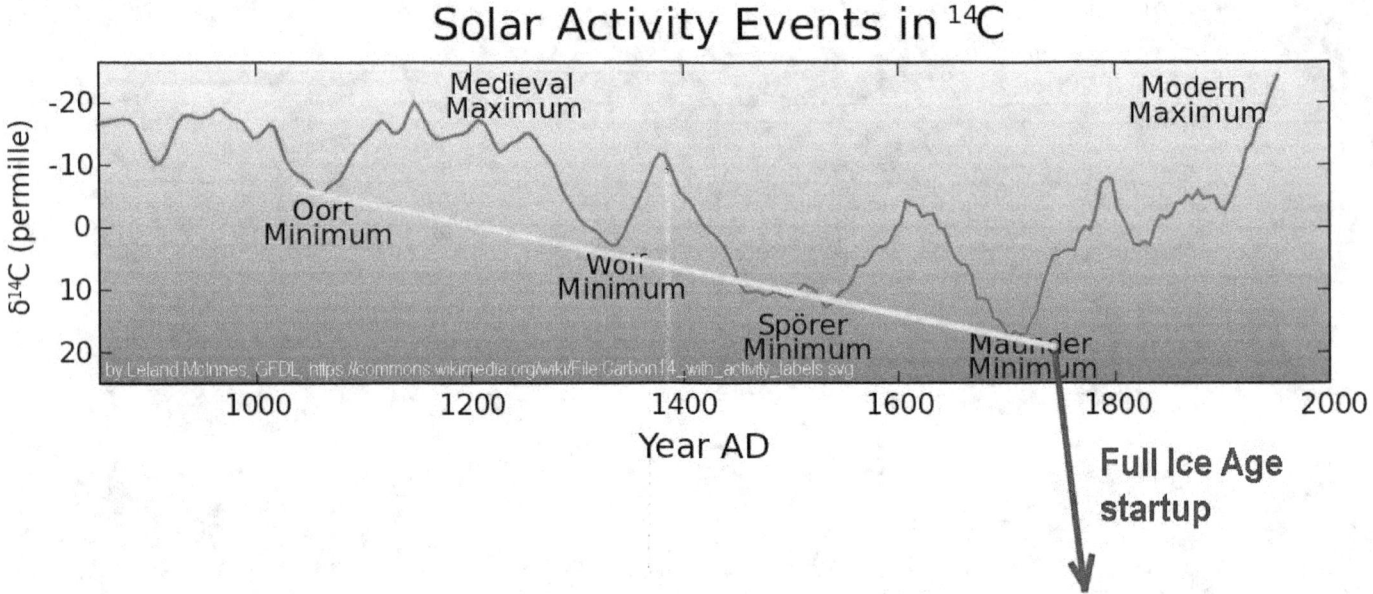

We didn't have this power developed in the 1600s. If the climate collapse had not been reversed at the time by the up-ramping of the Sun, but had continued, we would likely have crossed the threshold into the coming Ice Age that no one would have survived, or only a very few. None of us would exist today, had this happened. No one was prepared. The Ice Age phenomenon was unknown. Nor were the technologies developed that could have enabled society to avoid the consequences.

# Win, and come out richer with a brand new world

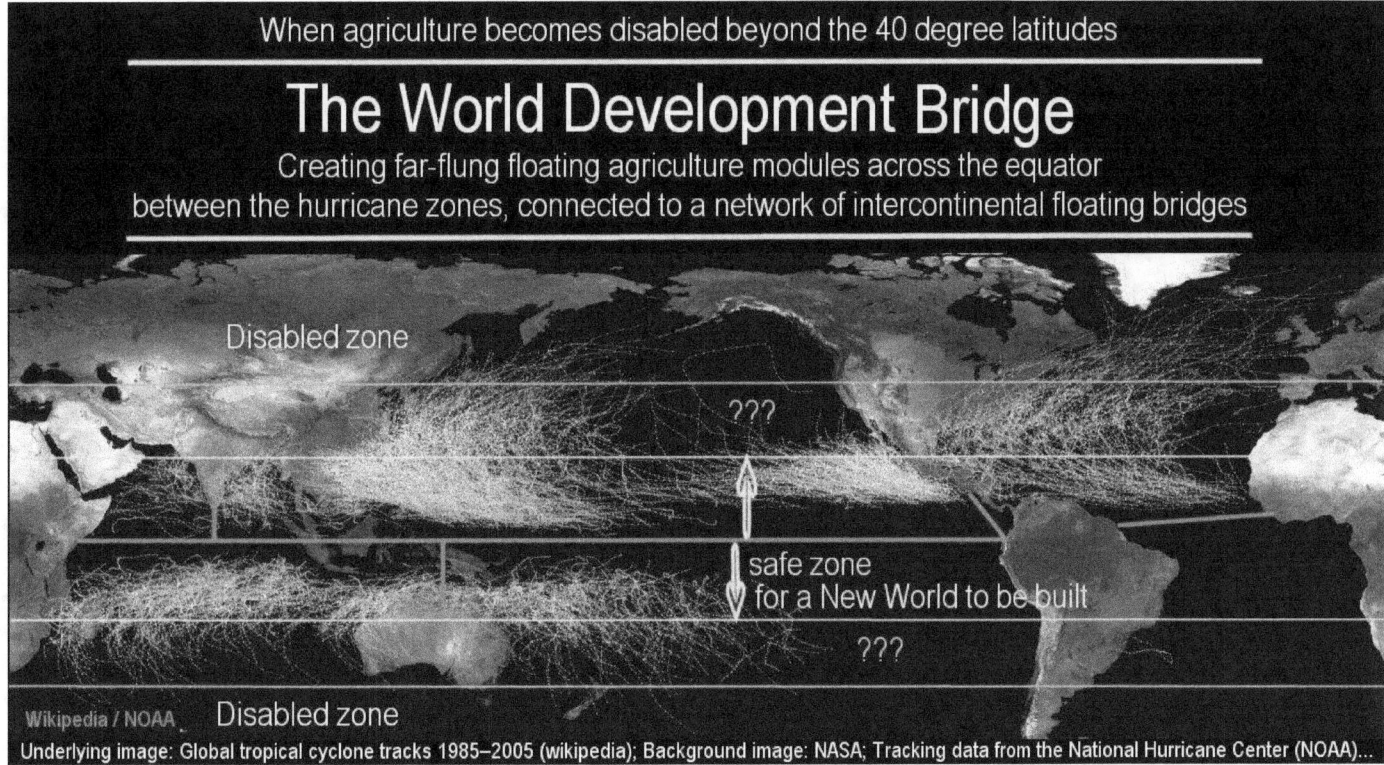

But today, we have the power at hand to meet the Ice Age Challenge and win, and come out richer with a brand new world. We have the power with technology to lift the death sentence that the Ice Age phase shift would impose, and enable us to continue to live. Doesn't this realization inspire Joy, and the Peace of Mind that comes with having a future, and the courage to experience the power that we have to make this future secure?

# We see the writing on the wall, and its promise

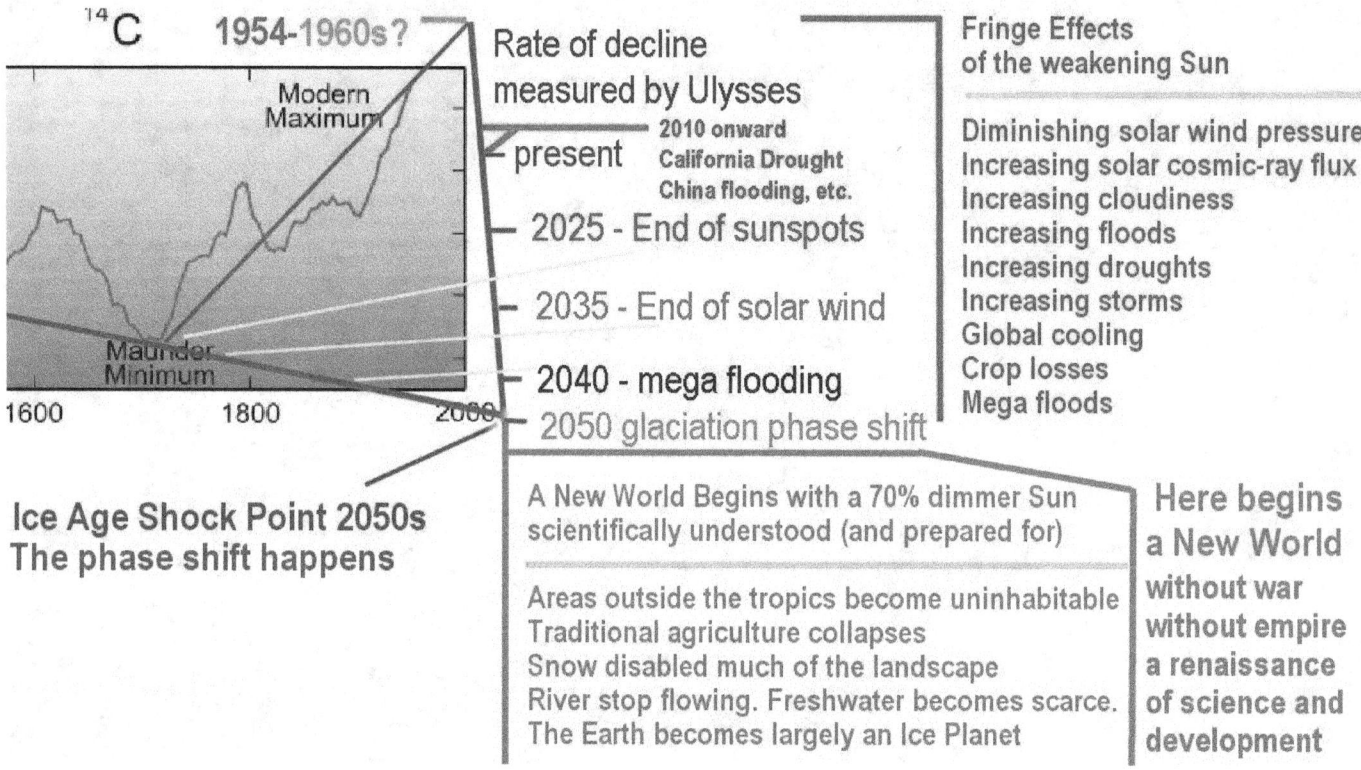

We see the writing on the wall, and its promise that when we get to the deep level of the diminishing background for the Sun, the resulting population collapse will become written in capital letters even before we will even get to the phase shift to the full Ice Age.

The writing on the wall tells us clearly, that this will be our destiny if we don't assert the power we have and build us new agricultures and new cities, and so forth, in the tropics, and get it done before the need for them becomes critical.

I would like to suggest that since the writing on the wall is boldly imperative, that what needs to be done to answer its call, will be done.

# There is joy in building, and no joy in being dead

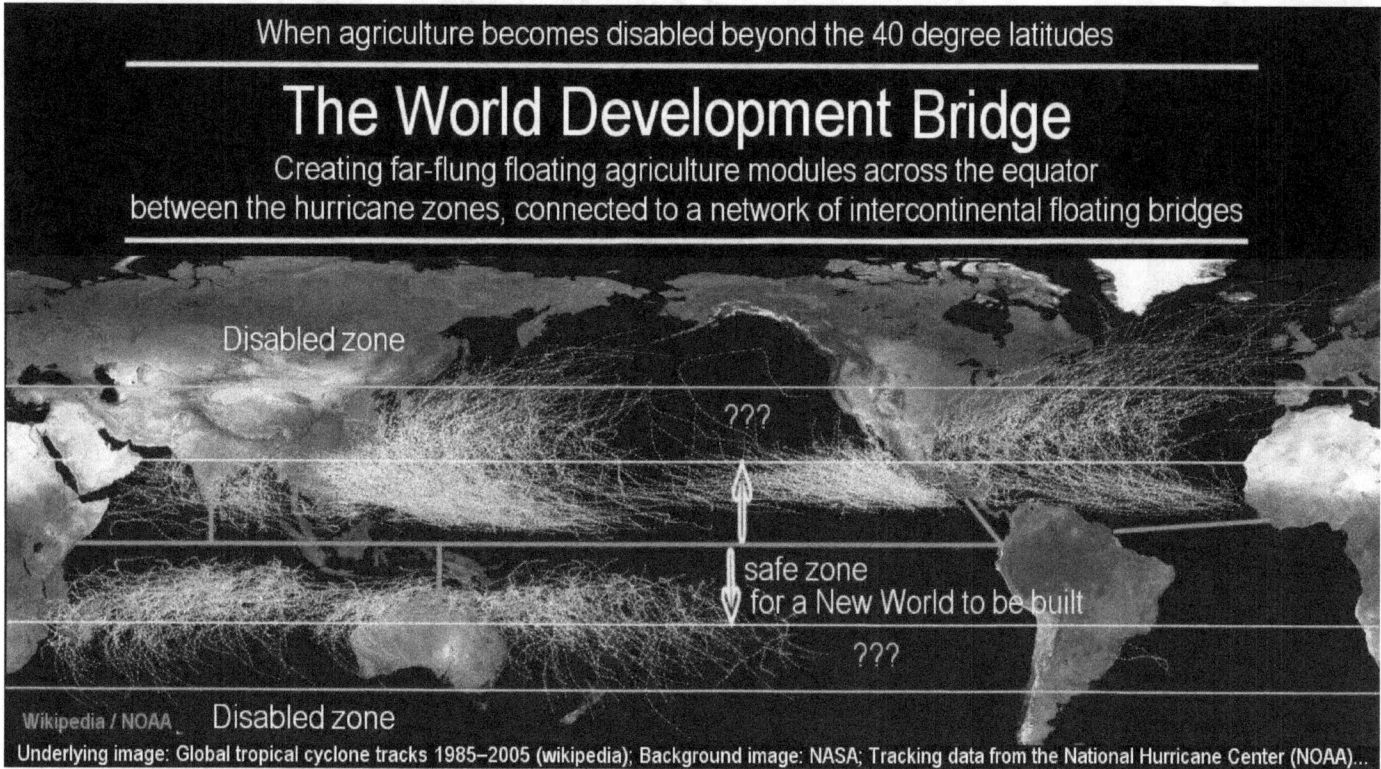

And it will be done as the greatest economic development adventure in the history of humanity. We have this option today. We can build us a new, high-technology world in the tropics that the climate collapse cannot affect, whereby the population collapse that is otherwise assured, can be avoided. We can build this New World afloat on the equatorial seas for the lack of suitable land in the tropics. We might build much of it with indoor agriculture and floating cities, creating an Ice Age infrastructure in the form of a World Bridge.

The materials, energy, and technology are all available. We only have to pick up what lays before us and run with it, and have the job completed before the population collapse crisis begins. The potentially gigantic population collapse can thereby be completely avoided.

I think we will go this route, because there is joy in building, and no joy in being dead.

# More Illustrated Science Books by Rolf A. F. Witzsche